AGENDA FOR NORTH-EAST INDIA

About the Editor

B. Datta Ray, a Tamrapatra awarded Freedom Fighter has authored two publications, Assam Secretariat 1874-1947 : An Administrative history of North East India and *Tribal Identity and Tension in North East India*. He has edited more than twelve volumes on North East India. He is the founder-secretary of North East India Council for Social Science Research since 1974, a pioneer and premier research body in North East India. He is the editor of the *Journal of North East India Council for Social Science Research*, a forward looking research Journal from North East India since 1977.

AGENDA FOR NORTH-EAST INDIA

Edited by

B. DATTA RAY

CONCEPT PUBLISHING COMPANY, NEW DELHI-110059

ISBN : 81-7022-938-3

First Published 2002

Published and Printed by
Ashok Kumar Mittal
Concept Publishing Company
A/15-16, Commercial Block, Mohan Garden,
New Delhi-110059

Phones : 5648039, 5649024
Fax : 091-11-5648053
Email : publishing@conceptpub.com

Dedicated to

Dr. S.P. Agrawal, Ph. D. (Honcausa)
Former Director
Indian Council of Social Science Research
Fighter for Freedom
and
Friend of North East India

ACKNOWLEDGEMENTS

This volume contains some of the papers which were presented to the workshop on the 21st century and the Emerging Scenerio and Issues of Social Science Research in North-East India, held in December 1999. It has been also enriched by the contributions of Professor Madhura Swaminathan, Professor Asok Kumar Maiti, Dr. S.B. Sinha and Dr. Amitava Mitra. The editor acknowledges their contribution and help.

In Professor J.B. Ganguly, the first Vice-Chancellor, Tripura University and in Professor Amalesh Banerjee of Rabindra Bharati University, Kolkata, I have discovered the perenial source of inspiration. In Professor Asok Kumar Maiti of Indian Statistical Institute, Kolkata, I found another true friend of researchers. I am grateful to all scholars whose article have been included in this volume.

I take this opportunity to thank Mr. Ashok Kumar Mittal of Concept Publishing Company, New Delhi for agreeing to publish Agenda expeditiously.

I acknowledge the technical support given by Ms. Kamala Chhetri and Mr. P.S. Thomas in preparing this volume. I would like to thank Professor P.M. Passah, President NEICSSR Professor B.J. Deb, Treasurer and Dr. S. Sarma Joint Secretary, NEICSSR for their help.

I would also like to thank Mr. M. Chakraborty of Voluntary Health Association of Meghalaya for his assistance in technical details.

30 March 2001 **B. Datta Ray**
Shillong

INTRODUCTION

North-East India with its large untapped hydel power, mineral resources, virgin forests, sturdy people and vast water resources remains in the backyard of development. These resources, should have changed the face of North East India with significant social and. economic development

In the 21st Century, the crucial challenge facing North East India is how to achieve a balance between economy and ecology in order to ensure sustainable social and economic development preserving the basic cultural matrix, ethical values and equaliterian ethos. This calls for the existence of truly democratic society that seeks to achieve economic efficiency and ecological harmony for the eradication of poverty and promotion of total human welfare. Such a situation can be built up by having an appropriate social and political structure that ensures the participation of men and women in real terms in the decision-making process at the different levels of the state and society. It is being increasingly realized that the pluralistic society has a significant role to play in counterbalancing the tendency to monolithic state absolutism. These are some crucial areas of social science research to which the papers included in this volume are directed. Researches on population, environment land food security are issues of vital interest of the problem of qualitative demographic development. Increasing number of elderly population would be an important component of the population. They should have a constructive role. We need to invest adequate quantum of resources for expansion on research, education and training facilities for human resource development in broader areas of social existence. For this we need the formulation of the appropriate development paradigm of North-East India which would ensure a meaningful and ecologically sustainable structural changes in the economy of North-East India, For this a systematic analysis and research of the total basic issues involving a balance between economy, ecology, population, human settlements patterns urban and rural areas physical resources, emerging scientific and technological knowledge application and resource management are required keeping in view the egalitarian ethos and social commitment.

Several aspects of human rights issues have surfaced. The new leadership of North East India is imbibed with the changing world view projecting the values of individual liberty and equality in the enjoyment of civil, political, social and economic rights. We are to reflect upon them based on empirical data.

Some of these papers were presented to the workshop on the 21st Century and the emerging scenario and issues of social science research in North East India held in December, 1999 in the occasion of the Silver Jubilee of North-East India Council for Social Science Research. Shillong. The papers included in this volume, depicts the emerging social milieu of North East India and suggests areas of research for the promotion of human welfare of the people in a strategically vital area known as North East India.

B. Datta Ray

ACKNOWLEDGEMENTS

This volume contains some of the papers which were presented to the workshop on the 21st century and the Emerging Scenerio and Issues of Social Science Research in North-East India, held in December 1999. It has been also enriched by the contributions of Professor Madhura Swaminathan, Professor Asok Kumar Maiti, Dr. S.B. Sinha and Dr. Amitava Mitra. The editor acknowledges their contribution and help.

In Professor J.B. Ganguly, the first Vice-Chancellor, Tripura University and in Professor Amalesh Banerjee of Rabindra Bharati University, Kolkata, I have discovered the perenial source of inspiration. In Professor Asok Kumar Maiti of Indian Statistical Institute, Kolkata, I found another true friend of researchers. I am grateful to all scholars whose article have been included in this volume.

I take this opportunity to thank Mr. Ashok Kumar Mittal of Concept Publishing Company, New Delhi for agreeing to publish Agenda expeditiously.

I acknowledge the technical support given by Ms. Kamala Chhetri and Mr. P.S. Thomas in preparing this volume. I would like to thank Professor P.M. Passah, President NEICSSR Professor B.J. Deb, Treasurer and Dr. S. Sarma Joint Secretary, NEICSSR for their help.

I would also like to thank Mr. M. Chakraborty of Voluntary Health Association of Meghalaya for his assistance in technical details.

30 March 2001 **B. Datta Ray**
Shillong

CONTENTS

LIST OF CONTRIBUTORS

S. K. Agnihotri, IAS C-II/78, Bapa Nagar, New Delhi-110003

K. Alam, Department of Economics, Guwahati University, Guwahati-781014

Meghali Baruah, Department of Education, Dibrugarh University, Dibrugarh-786004

K. Bez, Institute of Econometrics & Applied Statistics, Hegarabari, Guwahati-785036

M.P. Bezbaruah, Department of Economies, Guwahati University, Guwahati-781014

N.N. Bhattacharyya, Department of Geography, Guwahati University Guwahati-781014

Subhendu Chakraborti, Economics Research Unit, Indian Statistical Institute, Kolkata700035

S.B. Chakraborty, Anthropological Survey of India, Kolkata- 700016

Samir Kumar Das, Department of Political Science, Calcutta University, Kolkata-700027

K.K. Datta, I.C.A.R. Research Complex' Barapani-793103

Machumi Gogai, Department of Geography, Guwahati University, Guwahati

S.N. Goswami, I.C.A.R. Research Complex, Barapani-793103

Pranay Jyoti Goswami, Department of Economics, G.C. College, Silchar, Assam

A. S. Guha, Indira Gandhi National Open University, Shillong

Kalita, Department of Geography, Guwahati University, Guwahati

Biman Kar, Department of Economics, Guwahati University, Guwahati-781014

B.B. Kumar, Editor, *Dialogue*, New Delhi

Asok Kumar Maiti, Sociological Research Unit, Indian Statistical Institute, Kolkata-700035

V.S. Mahajan, Centre for Indian Development Studies, Chandigarh-160017

Xavier P. Mao, Department of Philosophy, North Eastern Hill University, Shillong-793014

Sukhendu Mazumdar, 29, Assam Rifles, 99 APO

Amitava Mitra, Department of Economics, Arunachal University Itanagar, Arunachal Pradesh

P.M. Passah, Department of Economics, North Eastern Hill University, Shillong-793014

Paramita Saha, Department of Economics, Tripura University Agartala-799004

B.K. Sarma, I.C.A.R. Research Complex, Barapani-793103

S.B. Singh, I.C.A.R. Research Complex, Barapani-793103

Kh. B. Singh, I.C.A.R. Research Complex, Barapani-793103

K.S. Singh, A-40/3, SFS Flats, Saket' New Delhi-17

Madhura Swaminathan, Social Science Unit, Indian Statistical. Institute, Kolkata- 700035

1

POLICIES OF FOOD SECURITY
An Agenda for Change

— Madhura Swaminathan

The situation in India today with respect to food stocks is unprecedented, and policies of public procurement, storage and distribution of food grain have been subjects of urgent discussion over the last few months. The Food Corporation of India (FCI) has over 40 million tonnes of grain in its stocks. At the same time, offtake from the public distribution system (PDS) is slackening, and large amounts of grain are rotting and going waste. Government policy in respect of the PDS has raised the prices of food grain and lowered the entitlements of poor households and has also excluded, on a very large scale, families from the category of BPL (below poverty line) from the reach of the PDS. Together, these policies have weakened the PDS and led to a decline in offtake. The rise in stocks means a rise in costs of storage — thus, the objective of reducing the food subsidy bill is not being achieved either. And all the while, of course, hundreds of millions of Indians remain hungry and malnourished. It doesn't take much wisdom to argue that present policies are inappropriate, and that major corrections in policy are needed.

Universal Coverage or the Case Against Narrow Targeting

The main argument for targeting is that of a reduction in fiscal expenditure. This, however, is a one-sided argument that focuses only on short-term effects. There are many costs of targeting, and when all costs are taken in to account, the case for targeting becomes tenuous. Given the specific features of the Indian situation, I argue that it is neither desirable nor feasible to implement a policy of narrow targeting. Instead we need a system of near-universal provision.

The Quantity Supplied and Adequacy of Ration Scale

If the system of food subsidy is to contribute significantly to household nutrition and food security, then a reasonable minimum quantity of grain needs to be distributed. The quantity distributed should bear some relation to minimum cereal requirements. According to the Indian Council of Medical Research (ICMR 1990), per capita cereal requirements amount to 135 kg a year (or 370 gms per person per day). The PDS should aim to provide, at the very least, about half the daily cereal requirements, that is 67 kg per person per year or 5.6 kg a month.

What would be the scale of the PDS if a ration of 70 kg, the achievement in Kerala, were to be provided to all hungry and undernourished households? If this ration were provided to the bottom 60 per cent (say 600 million) of the population, then the annual requirement of the PDS would be 42 million tonnes (or 25 per cent of net availability). In the 1990s, particularly after 1992, the quantity of food grain distributed in the PDS declined.

It is also important to ensure that the food grain supplied to the PDS conforms to certain minimum standards of quality.

Control over Food Prices

Since Independence, one of the key objectives of food policy and food subsidy programmes in India has been to maintain price stability and ensure that basic foods are available to the vulnerable population at reasonable prices. This goal was attained to the extent that the real price of basic cereals such as rice and wheat was kept in check, and declined through the 1970s and 1980s, and price fluctuations were lowered. Also, as shown by Professors Deepak Nayyar and Abhijit Sen, fluctuations in domestic prices have been much lower than fluctuations in world prices, and thus the consumer in India was protected from excessive price variations. In the 1990s, however, the trend has been reversed. The relative price of food commodities such as rice actually rose in the 1990s. In particular, the prices of rice and wheat in the PDS were raised repeatedly in the period 1992 to 1994. In a country where the majority of the population is not adequately nourished, it is essential that governments keep a check on inflation in food prices. The erosion of real incomes of the poor via high inflation in food prices has to be reversed.

It is not easy, of course, to establish, precisely what constitutes a "reasonable" or "affordable" price, but the price of cereals and other basic foods must bear some relation to wages, incomes and levels of poverty in society.

Production, Procurement and the Food Corporation of India

Since the PDS involves the distribution of food, it has to be supported by systems of procurement and storage. Policies of production and procurement have to be of a piece with policies of distribution, and specifically, an expanded PDS requires expansion of the production base and a more equitable system of procurement. We also need to examine the functioning of the Food Corporation of India (FCI), the body that implements the government's policy of procurement and storage.

It is alarming that the World Bank and others are arguing that it is now appropriate to "phase out government controls as well as procurement policies" (World Bank 1996: 40) on the grounds that India has moved from a situation of a deficit to that of surplus in grain production. This is a very short-run view. If all the hungry and undernourished consumed adequate amounts of food grain then present production would barely suffice to meet demand. One set of projections of cereal demand for 2020 show that 278 million tonnes may be demanded if per capita income grows at 3 per cent annually (Bhalla and Hazell 1997). In addition, if poverty were eliminated, cereal demand would rise to 292 million tonnes, and if the entire population were well fed, cereal demand would be 301 million tonnes *(ibid.)*. Clearly, domestic production needs to expand rapidly to meet this demand. In other words, for supply to keep pace with future increases in demand, it is necessary to expand the production base. In particular, there is need to raise productivity and production in more backward regions and, in dry-land areas, and such progress will require appropriate producer incentives.

For a country the size of India, the goal of self-sufficiency in domestic production, contrary to the logic of structural adjustment, cannot be abandoned. A reliance on imports to meet the food requirements of several hundred million persons will not only make Indian consumers vulnerable to world trade fluctuations but also undermine national sovereignty.

The food subsidy of the central government is the operational deficit of the FCI. The total economic costs of the FCI roughly doubled between 1990-91 and 1998-99 for its operations in both rice and wheat, resulting in a growing gap between costs and sales price. The nature of cost escalation showed that the fastest growing component of economic costs was the procurement price of food grain. As I argued above, we still need policies of procurement and price support but they must be viewed in conjunction with costs of the FCI and costs to the consumers via raised prices in the PDS. At the same time, it is obvious that costs of distribution (including freight, handling charges and overhead charges) have risen rapidly and these costs need to be controlled. There are also costs due to waste, specifically losses during transit and storage, and these must be brought down sharply.

We need to broaden the base of production and this should be accompanied by new and more diversified procurement. The operational costs of distribution and storage by the FCI must be kept in control by means such as greater transparency and accountability in day-to-day functioning. In short, there needs to be a unity of policy design with respect to procurement and distribution.

Administration and Functioning of the Food Delivery System

In most parts of the country, to be effective, the administration and management of the PDS need major overhaul. Much of the unevenness in the utilization of the PDS in different parts of the country and the associated failure to reach some of the poorest households and regions can be attributed to administrative failure. The food delivery network needs to be strengthened and administered in a non-corrupt and efficient manner for the poor to have access to the PDS. Some specific suggestions follow.

Employment and other Welfare Programmes

Some of the poorest households are unable to utilise the PDS because they lack the income to buy food. For such households, employment or other welfare programmes (such as pensions for senior citizens) are required to enable them to buy minimum quantities of food. In the discussion on "reform" of the PDS, it is often suggested that the resources withdrawn from the PDS be allocated to other welfare programmes such as employment programmes or nutrition programmes. I shall argue that employment and other social security programmes should be complementary to the PDS and not seen as a way of restricting food distribution only to households participating in employment or other welfare programmes or as substitutes for the PDS. Given the current level of stocks, food-for-work type employment programmes should be initiated with some urgency.

Conclusion

To sum up, I have outlined the main features of a strategy to strengthen the PDS. The first issue discussed was that of coverage. I argued in favour of universal or near-universal coverage on the grounds that the costs of targeting are high in a country such as ours where a large majority of the population is vulnerable to food deprivation. By the criterion of the food share, even with a high cut off level, a food share of 60 per cent, 90 per cent of rural households and 60 per cent of urban households would be eligible for programmes of food security. The conceptual case for broad and universal coverage of the PDS is fortified by the immense practical problems of administrative targeting based on criteria such as income poverty.

Secondly, to be effective, the quantity of food grain distributed in the PDS needs to be expanded. The ration scale should bear some relation to the norm for minimum cereal requirements. I argued that the PDS should attempt to provide about 70 kg of grain per annum — that is, the Kerala achievement and half the cereal requirements of a person. At the same time, the quality of food grain should conform to certain norms.

Thirdly, a check on the prices of basic foods is an essential component of any strategy of food security. The erosion of real incomes in recent years through high inflation in food prices must be controlled.

Fourthly, policies of production, procurement and storage have to be of a piece with policies of distribution. This requires continued emphasis on self-sufficiency in food production and an expansion of the production base. It also requires a revamping of the system of procurement with more diversified procurement. Lastly, the costs of operation of the FCI, the organisation responsible for procurement, storage and distribution, need to be lowered. At the same time, it should be recognised that a major component of the rise in costs, procurement price, is a variable outside the control of the FCI.

Fifthly, improvements in the administration of fair-price shops are imperative for the poor to have access to the PDS. Decentralisation of some functions to the local-level, to democratically elected panchayats, is possibly the right direction to go, and can help improve administration and monitoring of the working of the PDS.

Finally, a major factor limiting the access of the poor to the PDS is inadequacy of purchasing power. So programmes of employment generation and other welfare programmes are measures complementary to the PDS and need to be augmented.

With these measures, it is feasible to make the PDS an effective measure of food security.

2

GLOBALIZATION — YET ANOTHER THREAT TO FOOD SECURITY AND REGIONAL PARITY IN INDIA

—*Asok Kumar Maiti* and *Subhendu Chakrabarti*

Introduction

For a country with perennially food deficit situation, what was most important even in pre-independence period was availability of food. Kazi Nazar Ul Islam said" khudhatur Sishu chai na Swaraj chai Duto Bhat ektu noon" (Hungry child does not want freedom, it wants a few grains of rice and a little salt.). If we take the 72-years' period between 1878 and 1951, the population exactly doubled. Despite relatively low population growth rate during the colonial period, the food situation was far worse compared to that of the post-independence period when population growth rate was at the annual rate of 2.15 percent compound. The colonial domination was a more potent force than population growth rate as such for generation of poverty and destitution in India.

In the post-independence India the total production of food grains increased from 50.8 million tonnes (1950-51) to 185.1 million tonnes (1995-96). This was quite remarkable. The population of India increased from 360.0 millions (1951) to 846.3 millions (1991) i.e. around 2.5 times whereas the food grain production increased more than 3.5 times during the same period. Compared to the post-independence performance, the production in the pre-independence period of 72-years as mentioned before, remained almost stagnant. So there is every reason to feel proud of our achievement in the food front. The per capita availability of food grains was 510.1 grams per day in 1991 which indicates substantial production of food grains. At the time of writing this paper when our population has crossed 10 billion mark, the per capita per day (PCPD) availability of food grain is about 467.4 grams which is quite a deterioration in food production.

In spite of the above facts, the rural people below poverty line is 44.1 percent and the urban people below poverty line is 37.2 percent in 1998 thereby meaning that more than total population in 1947 can not yet afford two square meals a day. Greater production of food grains which enabled India to generate a huge marketable food surplus could not guarantee the food security of the masses of the people in India. Economists, especially the growth-economists who are happy with GDP figures should possibly look into the distribution system within the country rather than so-called efficiency of production.

Food security of a country basically means the physical and economic access of the entire population to the food grains. In other words, the population enjoying food security has at least adequate purchasing power to fulfil their minimum calorie needs. With communication and transport system inadequately developed and with the rising cost of transportation it is all the more necessary to have adequate production of food grains in each of the regions. Unfortunately there are still very backward regions in terms of communication system and other infrastructure facilities. North-Eastern states of India belong to one of such underdeveloped regions where food deficit (in terms of state domestic product) is perennial. Recent performances of these states in food front reveals a deteriorating situation in terms of availability of food grains per capita. Again, there are indications that the purchasing power of the population of North-East suffer much in terms of annual percentage change in the real wages for unskilled labour, most of whom are engaged in agriculture. So any decision on food security should not only address the per capita availability of food grain or even the employment as such but should also address the effective demand situation for food grains, the distribution system (both market and PDS), and also the logistics of supply situation.

The objective of PDS is to transfer resource to the needy through the supply of subsidised essential items. India Human Development Report (1999) presented data on estimated per capita expenditure as a percentage of total expenditure by different states. In the North-East region the food expenditure is 71.7 percent compared to the all India average of 63.9 percent. Expenditure on food grains alone is

Rs. 6.22 PCPD compared to the total expenditure of Rs. 8.67. This shows only the relative poverty of the population of this region. It is also alarming that the off-take of food grain under the PDS has been declining for several years. Whenever the off-take is less than procurement especially in a situation of high level of stocks, the operations unwittingly reduce overall market supply and push up prices. Food stock at the FCI therefore gives a false impression of abundance and this is primarily because of the fact that people's purchasing power has declined in recent years. Added to these is an important feature of price rise during recent period of food items compared to those of non-essential items.

The Alternative Economic Survey (1998-2000) argued conclusively that high production of food grains did not at all guarantee that food is available to those who need it. It was also estimated that about 300 millions of Indians do not have food security at all.

The PDS which is supposed to maintain the minimum nutritional status of the poor could not make much headway. In the past few years the importance of PDS as an outlet for food grains at controlled prices has diminished greatly as a result of frequent increase of procurement prices followed by more than proportionate increase in the central issue prices of food grains. The real beneficiaries of large increase in procurement prices have been the surplus farmers in only three or four states in India which account for nearly 90 percent of the procurement. The huge outlay in the name of food subsidy only helped sustaining food prices at a very high level.

It is true that the present PDS needs reforms so that the poor attains a kind of food security but in the recent times being pressurized by the WTO, IMF and World Bank the government is aiming at complete dismantling of the PDS. The union budget for the year 2000-2001 took decisive steps towards phasing out of the PDS. To the extent that the PDS provides food security to the poor is being seen as a hindrance to the free play of market forces in food sector. Notwithstanding many limitations of the system the PDS is a highly relevant scheme which need to be further strengthened in the country as a whole and more especially in the backward regions.

We would in the following paragraphs present the food situation in India with special reference to North-Eastern states and discuss the threat to food security in the context of globalization of our economy.

The Scenario in the Food Front

Ensuring food security and employment are two vital components of making any analysis of the food situation. The FAO document made some future projections for India regarding the demographic structure. The population in the country would be 115.22 crores in 2010 and 127.22 crores in 2020. The population dependent on agriculture which is at present 55.74 crores will touch around 57.84 crores in 2010 and then start declining to reach 57.47 crores in 2020.

Agriculture will still engage around half of the economically active population of the country in 2020 down from 61 per cent at present and 55 per cent in 2010. In a study (Chakraborty and Dutta in Alternative Economic Survey, 1998-2000) it was argued that agricultural activities will remain the mainstay of livelihood for the majority of the Indians for another 20 years and even at a conservative estimate assuming that the per capita availability of cereals at the prevalent rate 428.8 grams per day we would require at least 200 million tonnes. This implies that another 6 million hectares of land need be brought under cereal cultivation by the year 2020 at the prevailing rate of yield 1.85 tonnes per hectare. Alternatively, the yield rate should increase to 1.96 tonnes per hectare if sufficient land is not available. Either way, fulfillment of such requirement entails a 5.8 per cent growth over the next two decades. In this context it would be relevant to quote statistics of state-wise area, production and yield of food grains available from Department of Agriculture and Co-operation (vide Table 2.1). From this table read with table-2 providing estimation of total problem area of 1690.02 lakh hectares would show that even with the present yield rate such a target of 5.8 per cent growth would not be impossible to achieve. But, what we are witnessing in the current years is that there is an apparent shift of agricultural hectarage to non-agricultural use or uses away from cereal production. This has something to do with the policies of different state governments, especially in the wake of opening up of economy including our agricultural sector. We would refer back to Table-2.2 when we discuss the specific problems of the North-Eastern region. (see Table 2.1).

Table 2.1: Statewise Area, Production and Yield of Foodgrain

State	*1998-99*					*1997-98*					
	Area (M.hect.)	*%of Total area*	*Production (M. Tonnes)*	*% of Total Produ ction*	*Yield (KG/hect.)*	*Area (M.hect.)*	*% of Total area*	*Production (M.Tonnes)*	*% of Total Produ ction*	*Yield (KG/ hect.)*	*% coverage under irregation* (1995-96)
(1)	(2)	(3)	(4)	(5)	(6)	(7)	(8)	(9)	(10)	(11)	(12)
Andhra Pradesh	7.19	5.7	14.40	7.1	2003	6.52	5.3	10.82	5.6	1660	56.7
Assam	2.67	2.1	3.43	1.7	1288	2.72	2.2	3.58	1.9	1314	19.5
Bihar	8.96	7.1	12.91	6.4	1441	8.97	7.2	14.09	7.3	1571	47.3
Gujarat	3.90	3.1	5.57	2.7	1426	4.08	3.3	5.71	3.0	1400	29.3
Haryana	4.49	3.6	12.12	6.0	2700	4.18	3.4	11.35	5.9	2715	77.3
Himachal Pradesh	0.84	0.7	1.49	0.7	1760	0.85	0.7	1.44	0.7	1691	18.5
J&K	0.88	0.7	1.52	0.7	1728	0.89	0.7	1.42	0.7	1587	38.8
Karnataka	7.38	5.9	9.98	4.9	1352	7.06	5.7	8.05	4.2	1140	21.4
Kerala	0.39	0.3	0.69	0.3	1768	0.43	0.3	0.80	0.4	1855	51.4
Madhya Pradesh	17.79	14.2	19.80	9.8	1113	17.86	14.4	17.36	9.0	972	30.8
Maharashtra	13.09	10.4	12.75	6.3	974	13.18	10.6	9.66	5.0	733	12.7
Orissa	5.38	4.3	5.81	2.9	1080	5.48	4.4	6.64	3.5	1211	28.1
Punjab	6.12	4.9	22.91	11.3	3741	5.88	4.7	21.14	11.0	3596	96.8
Rajasthan	13.46	10.7	13.93	6.4	961	13.75	11.1	14.05	7.3	1022	23.6
TamilNadu	4.45	3.5	10.14	5.0	2278	3.64	2.9	8.10	4.2	2225	60.4
Uttar Pradesh	20.52	16.4	40.15	19.8	1957	20.50	16.6	41.59	21.6	2029	65.0
West Bengal	6.54	5.2	14.37	7.1	2198	6.56	5.3	14.35	7.5	2189	28.0

Source : Department of Agriculture & Cooperation Statistics at a Glance, 1995

Table 2.2 : State-wise Estimated Total Problem Area, Area Treated and Balance Area yet to be Treated

(Area in lakh hector)

State	*Total Problem Soil erosion*	*Area due to Land degradation*	*Total Problem*	*Area Treated till Area end 1993-94*	*Balance yet to be Treated*
(1)	(2)	(3)	(4)	(5)	(6)
Arunachal Pradesh	115.02	7.29	122.31	13.59	108.72
Assam	22.17	7.82	29.99	2.30	27.69
Manipur	3.74	3.60	7.34	1.16	6.18
Meghalaya	8.37	2.65	11.02	1.23	9.79
Mizoram	4.21	1.89	6.10	0.22	5.88
Nagaland	4.05	6.33	10.38	1.06	9.32
Tripura	1.67	1.12	2.79	1.59	1.20
North-East	159.23	30.70	189.93	21.15	168.78
India	1234.47	455.55	1690.02*	373.60	1316.42*

* Coastal Sandy-Area and Acid Soil is not included

Source : Indian Agriculture in brief. 26th edition, 1995. Dept. of Agriculture and Cooperation, Govt. of India

A recent study by IIM Ahmedabad commissioned by ministry of agriculture, government of India showed that output increase in food grain sector is much more influenced by non-price factors, like electricity, rainfall and technical change than the change in prices with the institutional factors of farmers' speed of adjustment being the least influential. It was also found by them that the marketed surplus response of all commodity groups, namely food grains and allied agricultural products, major non-food grains and all agricultural products are highly price elastic. The study implies that pure market oriented strategy that relies completely on price signals to arrive at economic decisions will not be very effective in either increasing the hectarage or the supply of food grains in general and cereals in particular. So pure market oriented strategy in food front would not raise food grain output and an utmost care is called for ensuring necessary investment in agriculture both in the public and the private sectors. The present day apathy of the state in opting for sustained gross capital formation would prove counter productive in terms of national food security.

Table 2.3 below presents the annual growth rates of production of food grains in the country. It shows sharp fluctuation in both total cereals as well as in total food grains. It is also alarming that the growth of coarse cereals which are common peoples' food has been declining in the recent decades. Something need be done in order to see that agricultural acreage do not go away from cereals in general and coarse cereals in particular in order that food prices do not go beyond the reach of the common people.

Table 2.3 : Annual Growth Rate of Production of Foodgrains (Index Based) (Base triennium ending 1981-82 = 100)

(Percent per annum)

Crop	*1967-68 to 1979-80*	*1979-80 to 1989-90*	*1989-90 to 1998-99**
(1)	*(2)*	*(3)*	*(4)*
Rice	1.99	4.29	1.60
Wheat	5.68	4.24	3.62
Coarse Cereals	0.67	0.74	-0.48
Total Cereals	2.47	3.63	1.88
Pulses.	-0.44	2.78	1.19
Foodgrain	2.02	3.54	1.80

* Provisional

Note : Exponential function using least square method has been used to obtain the compound growth rate

Source: A pre budget 2000-2001 document, Indian Economic Survey- 1999-2000

Table 2.4 : Per Capita Availability of Foodgrains in India

(Grams per day)

Year	*Cereals*	*Pulses*	*Foodgrains*	*Decadal change*		
(1)	*(2)*	*(3)*	*(4)*	*(5)*		
1951	334.2	60.7	394.9			
1961	399.7	69.0	468.7			
1971	417.6	51.2	468.8			
1981	417.3	37.5	454.8	Decade	Mean	S.D
1991	468.5	41.6	510.1	1951-60:	429.8	26.8
1992	434.5	34.3	468.8	1961-70:	447.5	23.7
1993	427.9	36.2	464.1	1971-80:	442.2	25.4
1994	434.0	37.2	471.2	1981-90:	464.2	16.4
1995	460.6	38.1	498.7	1991-99:	480.1	20.0
1996	447.0	33.2	480.2			
1997(P)	471.8	37.5	509.3			
1998(P)	417.5	33.2	450.7			
1999(P)	428.8	38.6	467.4			

(P) = Provisional

Source : Dept. of Agriculture and Cooperation Statistics at a Glance

As mentioned earlier, the per capita availability of cereals per day for the country as a whole is 428.8 grams in 1999 whereas the figure of total food availability is 467.4 grams (vide Table 2.4). If we look into the decadal growth of food grains availability per capita per day (PCPD) we find a steady increase from 394.9 grams in 1951 to 467.4 grams in 1999, which is commendable. It does not however show that we are producing enough food if we want to fulfil the total need of the entire population. The minimum need if calculated on the basis of 480 grams Per Capita Per Day (PCPD) for urban and 540 grams PCPD for rural population, we have still deficit in production.

The progress above took place inspite of the fact that not much was done for improvement of cultivable land. Even after 50 years of planned economic development around 60 percent of our net sown area is mono-cropped. Irrigation (both surface and under ground) caters to only 47.02 per cent of the area under cereals and 12.67 per cent area under pulses (vide Table 2.5). The abysmal condition would be much more pronounced if we consider India with exception of four highly agriculturally developed states. The picture is bleak if one considers the situations in the backward regions.

Table 2.5 : Percentage of Irrigated Area to Total Area

State	*Total Cereals*	*Total Pulses*	*Total Foodgrains*
Assam	20.11	5.00	19.46
Arunachal Pradesh	19.41	—	18.64
Manipur	43.35	—	43.35
Meghalaya	34.88	—	34.09
Mizoram	6.85	—	6.49
Nagaland	34.04	—	32.00
Tripura	17.49	20.00	17.58
All India	47.02	12.69	40.64

Source : Ibid

The situation depicted above shows that availability of food for all could not be achieved even in 2000, judged from the need of the people. Since around 300 millions of Indians could not generate demand for minimum food requirement, the country can export food and there are sufficient stocks of food grains. On the other hand India has begun importing food grains from abroad under the compulsion imposed by WTO of which India is a member. Moreover, the government is insistent on soft-peddling the PDS while more enthusiastic among them are in favour of abandoning the system altogether. Such development is a real threat to the national food security system and much more dangerous than the requirement production gap of food situation.

It would be interesting to note that of the 308 items of agricultural sector's export during 80's, only 120 items were primary products. India stopped exporting 26 odd items in the post 1991 period, most of which are processed products. In the post 1991 period there has been a sharp change in the composition of exports in favour of primary products at the cost of sharp decline in processed products. From the above it is also clear that quality provisions in WTO are putting up non-tariff barriers to our exports. Moreover, provisions for compulsory imports of agricultural products is creating an added threat to our agricultural system as such.

Judging the overall food-security situation in our country, especially after adoption of liberalization as an economic policy, we find that the security of the vast majority of Indian population (most of whom are engaged in agricultural sector) is in serious jeopardy. Quantitative restriction on import of agricultural commodities were removed under the WTO obligation and subsidies on inputs necessary for agricultural production are being drastically cut. Prices of food grains offered under PDS have been hiked ostensibly to honour WTO commitment, whereas quality clause put embargo on Indian exports. The employment status of the population engaged in agricultural sector has substantially downgraded. All these facts pointed out that India in no way would benefit under globalization and more particularly the backward regions wholly or mainly dependent on agriculture would be heading towards a serious crisis if the government fails to reverse the new agricultural policy.

Let us now look at the regional situation in North-East regarding the need-performance syndrome of agricultural products. Table 2.6 showed that of the total problem area of 189.93 Lakh hectares in the North-Eastern states, only 21.15 Lakh hectares were treated. So a vast area having potential of foodgrain production has not yet been explored at all. The potential of foodgrain production in the uncultivated but cultivable land is more than sufficient to satisfy

Table 2.6 : Annual Average Growth Rate of Population and Production of Foodgrains in the N-E States and All India

State	*1980-90*		*1991-99*	
	Population	*Production*	*Population*	*Production*
Assam	1.5	1.3	1.2	1.5
Arunachal Pradesh	3.1	4.2	3.1	-1.8
Manipur	2.6	-1.7	2.5	2.2
Meghalaya	2.8	-0.4	2.8	2.0
Mizoram	3.4	8.3	3.3	7.4
Nagaland	4.3	3.7	4.4	4.5
Tripura	3.4	1.2	2.9	0.2
All India	2.1	1.8	1.9	1.8

* 1983-90 & 1991-99

Source : Census of India - 1981, 1991; Economic Survey-various issues.

the needs of the entire population of the North-Eastern states even taking the growth of population into account.

Table-2.6 shows the annual average growth rate of population and production of foodgrains. Except in Arunachal Pradesh the overall performance of all other states are encouraging. What we plead here is that if the treatment of yet untreated area of 168.78 lakh hectares is done, North-Eastern region can over-fulfill its food need and get rid of its perennial food shortage. Unfortunately, no such specific policies are being pursued by any of the state Governments of this region. One wonders when these Governments put so much emphasis on problems of insurgency and other things how they are so indifferent to the causes of such insurgencies. Many scholars have pointed out that there are strong linkages between economic maladies and insurgency problems. Food system is therefore linked with politics in an indirect way. The state governments can possibly deal with their law and order problems better if they take care of employment and build up of food security of the entire population in this region.

Tables 2.7a, 2.7b and 2.7c present the performance of North-East states vis-a-vis all India regarding availability of cereals (Table 2.7a), of pulses (Table 2.7b) and total foodgrains (Table 2.7c) per capita per day. The time series data of 17 odd years from 1983 to 1999 show that there has been general improvements in performance of the states except for Meghalaya. So judged from the performance of the primary producers the situation is optimistic. But the real

Table 2.7a: Availability of Cereals Per Capita Per Day

(in Grams)

Year	*Assam*	*Manipur*	*Meghalaya*	*Nagaland*	*Mizoram*	*Tripura*	*Arunachal Pradesh*	*All India*
(1)	*(2)*	*(3)*	*(4)*	*(5)*	*(6)*	*(7)*	*(8)*	*(9)*
1983	365.78	418.50	293.75	392.64	179.05	535.55	569.75	448.41
1984	353.17	473.94	303.60	330.38	150.58	469.29	858.43	520.27
1985	342.79	600.58	287.59	344.98	226.69	446.29	886.62	487.69
1986	384.51	588.57	292.53	310.15	232.85	427.56	800.23	489.97
1987	324.02	428.65	213.77	246.57	237.46	433.25	671.78	461.03
1988	359.54	461.00	235.18	247.23	233.16	475.81	675.42	443.56
1989	322.62	451.82	222.22	369.67	272.64	488.65	666.04	424.14
1990	358.73	397.83	236.01	402.60	280.51	476.60	684.71	520.50
1991	414.77	425.91	234.64	411.57	297.50	506.57	679.10	521.52
1992	402.07	508.86	229.42	408.59	319.99	463.48	661.21	493.64
1993	405.31	392.97	208.26	458.35	360.00	420.51	562.82	416.60
1994	410.19	492.35	208.24	432.05	398.46	457.13	653.90	520.47
1995	399.72	657.63	193.73	401.10	399.26	371.43	524.41	530.77
1996	403.54	444.48	188.03	411.56	392.19	405.14	535.85	494.17
1997	394.16	499.42	228.85	344.56	412.77	458.31	522.72	535.21
1998	395.05	544.49	232.94	372.18	398.57	438.12	517.02	505.90
1999	373.89	476.35	225.94	423.55	395.80	391.87	446.56	522.71

Source: Census of India - 1981, 1991; Economic Survey-various issues.

Tabl 2.7b: Availability of Rules Per Capita Per Day

(in Grams)

Year	*Assam*	*Manipur*	*Meghalaya*	*Nagaland*	*Mizorarn*	*Tripura*	*Arunachal Pradesh*	*All India*
(1)	*(2)*	*(3)*	*(4)*	*(5)*	*(6)*	*(7)*	*(8)*	*(9)*
1983	7.10	2.93	3.88	3.88	0.52	3.02	—	45.19
1984	6.72	—	3.96	8.66	3.51	3.05	—	48.09
1985	7.81	—	5.13	21.30	1.46	2.96	—	43.68
1986	8.59	—	4.09	23.20	1.41	2.88	—	47.76
1987	7.76	—	4.67	5.96	2.27	2.91	—	40.98
1988	7.71	—	4.03	9.58	4.39	4.13	—	37.58
1989	6.40	—	3.76	13.62	4.67	4.74	—	46.51
1990	6.31	—	3.81	21.79	2.87	6.24	—	42.31
1991	5.93	—	3.70	27.18	26.22	6.16	13.31	45.89
1992	6.44	—	3.60	32.50	26.51	6.08	13.82	37.93
1993	6.10	—	3.50	15.54	29.35	6.09	15.77	39.72
1994	6.72	—	3.54	19.82	35.21	5.91	15.29	40.51
1995	6.92	—	3.31	14.98	34.40	5.12	15.65	41.99
1996	6.58	—	3.35	21.39	23.52	3.94	15.44	36.19
1997	7.78	—	3.25	23.58	22.10	4.66	17.59	41.17
1998	7.25	—	3.16	20.73	21.06	4.37	16.79	36.63
1999	7.69	—	3.07	22.051	28.581	3.37	16.77	41.14

Source : Census of India - 1981, 1991; Economic Survey-various issues.

Table 2.7c: Availability of Foodgrains Per Capita Per Day

(in Grams)

Year	*Assam*	*Manipur*	*Meghalaya*	*Nagaland*	*Mizorarn*	*Tripura*	*Arunachal Pradesh*	*All India*
(1)	*(2)*	*(3)*	*(4)*	*(5)*	*(6)*	*(7)*	*(8)*	*(9)*
1983	372.88	421.44	297.63	396.52	179.57	538.56	569.75	493.60
1984	359.91	473.94	307.56	339.04	154.09	472.35	858.43	568.36
1985	350.61	600.58	292.72	366.29	228.15	449.25	886.62	531.36
1986	393.16	588.57	296.62	333.36	234.26	430.44	700.23	537.72
1987	331.77	428.65	227.08	252.53	239.73	436.16	671.78	502.01
1988	367.25	461.00	239.21	256.81	237.55	479.93	675.42	481.14
1989	329.02	451.82	225.98	383.29	277.31	493.39	666.06	570.65
1990	365.04	397.83	239.82	424.38	283.39	482.84	684.71	562.81
1991	420.69	425.91	238.35	438.75	323.72	512.73	692.41	567.41
1992	408.51	508.86	233.02	441-09	346.50	469.56	675.03	531.57
1993	411.41	392.97	211.76	473.89	389.35	426.60	578.59	556.32
1994	416.92	492.35	211.78	451.87	433.68	463.04	669.18	560.98
1995	406.64	657.63	297.03	416.08	433.66	376.55	540.06	572.76
1996	410.14	444.48	191.38	432.95	415.71	409.09	551.29	530.35
1997	401.95	499.42	232.10	468.15	434.87	462.97	540.31	576.37
1998	420.30	454.49	236.10	392.91	419.63	442.48	533.81	542.52
1999	.381.58	476.35	228.99	445.60	424.38	395.25	463.33	563.84

Source: Census of India — 1981, 199 1; Economic Survey-various issues.

wage situation of agricultural labourers is very poor. Table 2.8 presents the annual percentage change of real wages for agricultural labourers in selected states compared to the state of Assam in the North-East region (no state-wise figures for other North-Eastern states are available yet). The decline in real wage, inspite of the growth

Table 2.8 : Annual Percentage Change in Real Wages for Agricultural Labour for Selected States

[Percentage change for agricultural year (July to June) over previous year]

State	*1992-93*	*1993-94*	*1994-95*	*1995-96*	*1996-97(p)*	*1997-98(p)*	*1998-99(p)*
Assam	+0.58	–6.58	–1.67	+2.68	+1.52	+0.65	–3.89
Bihar	–5.00	+5.98	+1.69	–2.30	+15.15	–4.70	–5.70
Gujrat	+7.92	+2.86	+1.27	+2.92	+5.08	+14.43	+13.41
Punjab	+4.25	+1.51	–1.17	–6.50	–0.42	+0.68	–3.03
U.P	+7.56	–6.71	–2.31	+14.78	–6.39	+15.92	+1.62
W.Bengal	+24.39	–6.50	–5.29	–0.28	+11.15	+3.02	–9.90
All India	+5.21	+5.61	–0.39	+0.72	+6.37	+7.11	–2.12

Source : A pre budget 2000-2001 document, Indian Economic Survey, 1999-2000

of production is a matter of concern not only from the point of food security but also from the points of social and political insecurity.

With the adoption of New Agricultural Policy in the monsoon session (2000) in the Indian parliament, there has been removal of quantitative restrictions on imports of around 1400 agricultural commodities. This sudden opening up of India's agricultural sector would threaten the food security system in its lock, stock and barrel. The experiences of Phillipines, China and other such developing nations notwithstanding, India allowed its economy as a dumping ground of genetically engineered cheap food by the chemical giants, the MNCs and TNCs. So it is apprehended that there would be downward adjustment of real wages and massive unemployment in the agricultural sector. Unless these policies are revised there is little hope that there would be any food security for the overwhelming majority of the Indian rural poor.

The Challenge Ahead

At the beginning of the new millennium we are facing a serious challenge related to our very existence. In 1947 when we began our 'tryst with destiny' we were poor but hopeful, dreaming of a self-reliant, self-sufficient, sovereign country which would be built up by us. After 53 years of independence we have achieved no less. But our government has sacrificed the basic principles of development and reduced the primary producers to a hopeless situation where eking out a living is left at the mercy of foreign capital. Must we, the people of India, meekly submit to the dictates of the Trio (IMF, WB and WTO) and sacrifice our principle of self-dependence and economic sovereignty or stand up to defend our independence with which the lackeys of foreign capital are toying? Whose millennium would this be? Ours or theirs?

REFERENCES

1. Statistics at a glance, Department of Agriculture and Cooperation, Government of India, 1995.
2. Indian Agriculture in brief, 26th edition, Department of Agriculture and Cooperation, Government of India, 1995.
3. A pre budget 2000-2001 document, *Indian Economic Survey*, 1999-2000.
4. Alternative economic survey- 1998-2000, Alternative survey group, Rainbow Publishers Ltd. June-2000.
5. India Human Development Report, 1999.

3

FOOD SECURITY AND AGRICULTURAL GROWTH

Prospects for the North-Eastern Region

— M. P. Bezbaruah

The term food security, as understood today, implies both physical and, economic access to balanced diet for each household and for all members in a household. As a report of the Food and Agriculture Organisation of the UN puts it, food security implies that 'all people at all times have both physical and economic access to the basic food they need' (FAO 1983, pp. 33). World Bank (1986) report is even more specific. It states that food security means 'access by all people at all times to enough food for an active, healthy life. Its essential elements are availability of food and the ability to acquire it'. The bank has also drawn a distinction between chronic and transitory food insecurity. Chronic food insecurity reflects 'continuous inadequate diet caused by the inability to acquire food. It affects households that persistently lack the ability to either buy food or produce their own.' The transitory food insecurity is defined as temporary decline in the household's access to food. It results from instability in food price, food production and household income—and in its worst from. It production and household income form, it produces famine.'

Food security is thus concerned with both production and distribution of food. To achieve sustainable global food security, first of all there must be enough current and future food production to meet world population's need for adequate nutrition. Given that there is adequate global production, the different countries of the world should be able to acquire their required supply either from their own share in the world production, or through trade. Given that there is enough national supply, the households in the country must enjoy the entitlement for their minimum food requirement. This entitlement will depend on a household' s own resource position, the way it decides to use its resources, as well as on external factors such as the price of food and the value its own resources can fetch from exchange (Sen, 1981, Ch. 1).

Adequate world food production on itself therefore is no guarantee for food security even at the national aggregate level, leaving alone household and individual levels. A country that is not self sufficient in food production may not enjoy enough entitlement for its food security if it lacks the economic and political strength to acquire its' requirement through exchange from the surplus nations. Some measure of self-sufficiency in food production may therefore be a justifiable national goal irrespective of sufficiency of world production. In fact countries cutting across levels of economic development are so cautious about food security, that even developed market economies are found to be bearing the burden of subsidization and high consumer prices to protect their uncompetitive agriculture.

Thanks to the Green Revolution or the HYV seed bases technology, many developing countries including India have been able to achieve sustained increase in domestic food production over the last three decades or so, thereby securing for themselves enhanced entitlement of food at least at the national aggregate level. (That the price of basic food items such as cereals had to be kept high or that the purchasing power of the poor could not be enhanced sufficiently to distribute the aggregate national entitlement equitably is a different matter.) Given this self-sufficiency in production of basic food at the national level, should the different regions of the country strive for regional self-sufficiency? So long as there is no real restrictions - physical, economic or legal, on the movement of food across the length and breath of the country, the answer in general is 'no'. Given that the 'goal of adequate national food supply is fulfilled, to maximise GDP the different regions of a country should in principle diversify and specialise in lines of production in which they enjoy comparative advantage. However India's North-Eastern region stands on a different footing in this regard and a strong case for self-sufficiency in food production for enhancing regional food security can be made out. Its narrow road and rail links with the rest of the country being prone to disruption from both natural and man made conditions, dependence on other parts of India for bulky essential supplies like food-grains can at times prove to be an unsafe proposition. Even without the disruption of links, the high transport cost of bulky imports into the region adds significantly to the social cost of maintaining regional food security without regional self-sufficiency in the production of at least the staple food-grains. Further it is now widely recognised that the region possesses the key resources of land and water in abundance for achieving higher agricultural production in general and self-sufficiency in production of its staple cereal rice.

It may however be noted that all parts of the region are' neither suited nor required to concentrate in production of food-grains. In plain areas, which are better suited for the HYV seed - fertilizer technology, intensive use of improved inputs with better water management can greatly enhance production of foodgrains.[1] Hills are however better suited for horticultural and plantation crops. In the absence of easy access to market network many households in the hills still depend on traditional *jhum* cultivation for subsistence. The system of *jhum* cultivation may not have endowed the people with great riches, but the system traditionally ensured a minimum food security in most times to the people living the remote hills. But with *jhum* cultivation gradually losing sustainability, the food security associated with it is also getting endangered. On the other hand, with the changing conditions,.especially in communication, alternatives to *jhum* cultivation are-also gaining viability. Finding the new alternatives profitable, a section of hill population traditionally dependent on *jhum* cultivation has already adopted cultivation of commercial crops, especially in areas in which better communication links and market access has been established.[2] But for the hill farmer to make a complete and successful transition from a subsistence *jhum* cultivator to a market force guided commercial farmer connectivity has to improve even further, so that markets get integrated and Public Distribution System (PDS) or some alternative network of distribution of essential supplies reach out to areas now considered too interior. Without such developments however it will be unrealistic to expect *jhum* cultivation to phase out quickly. In the absence of improved communications, farmers would obviously not have the market outlets to sell their produce, nor would basic food stuff be available to be procured in exchange. Giving up subsistence farming for growing commercial crops would then jeopardise whatever food security the *jhumia* communities currently enjoy.

The government efforts therefore need to be concentrated in organising the necessary infrastructure support in the areas of road communication, irrigation, storage and PDS. As such supports come up, the farmers in the hills and plains can specialise in mutually complementary cropping patterns securing for the whole region not merely food security but general agricultural prosperity and a widely shared broad based pattern of economic growth.

NOTES

1. The recent spurt in rice production in the Brahmaputra Valley of Assam following a crash programme of installation of one hundred thousand shallow tube wells for irrigation confirms such belief.
2. This observation is based on the author's findings in Karbi Anglong district of Assam during a research project entitled 'Towards a Strategy for Economic Development of Karbi Anglong'. The author was an Advisor in the Reform Committee on Economic Affairs of the Karbi Anglong Autonomous Council which carried out the project and submitted the report in June 2000.

REFERENCES

FAO (1983), Report on Eight Session of the Committee on World Food Security, Rome

Sen, A. K. (1981), *Poverty and Famine, An Essay on Entitlement and Deprivation*, Clarendon Press, Oxford,

World Bank (1986), *Poverty and Hunger. Issues and Options for Food Security*, Washington D C

4

FOOD INSECURITY IN INDIA
With Special Reference to North Eastern Region

— *P.M. Passah*

The study made recently by the United Nations Population Fund (UNFPA)[1] has revealed that millions of the rural poor in India are not adequately fed despite the sizeable gains achieved in agricultural productivity. The rural poor are being denied their daily full quantum of energy requirement in contrast to the rapid rise in urban prosperity. One-third or around 330 million of India's population in the country side suffer from food insecurity. This means that the higher economic growth since 1994 has not been able to reduce the number of absolute poor. Thus drastic action has to be taken on the agricultural sector along with properly designed poverty alleviation programmes so as to guarantee food security to the rural poor in the future. It is important to identify the reasons for failure on the food front for the poor notwithstanding the strong and well-manned Public Distribution System in place.

Agricultural Achievement

As a result of the Green Revolution, the total foodgrain production in India rose from a mere 50.8 million tonnes in 1950-51 to 192.43 million tonnes in 1997-98, a four-fold increase resulting in a buffer stock of 35 million tonnes. This was possible because of the increase in productivity from a mere 644 kgs. per hectare in 1949-50 to the current 1551 kgs. per hectare. This increase was clearly manifest in the case of rice (1895 kgs. per hectare) and wheat (2470 kgs. per hectare). However, the per capita net availability of staple food according to the Indian Council of Agricultural Research (ICAR)[2] rose only to 216.4 grams and 168.2 grams per day in 1998 which works out to about 79 kgs. and 61.4 kgs. per year in the case of rice and wheat respectively.

Stagnation in Foodgrain Production

Though such spectacular growth has been achieved, yet in the last few years the total foodgrain production stagnated at around 198 million tonnes. According to the UNFPA, per capita foodgrain production in India is a meagre 200 kgs. per year, just a little above the UN Standard for famine-like conditions which is below 200 kgs. per person per year.

The *16.5 million addition to the population* every year requires, according to the UNFPA estimate, the raising of foodgrain production from 4 million tonnes to 5 million tonnes annually. But the country will now have to achieve the targetted food production from less land with less water, expensive labour and at higher cost of other inputs. *A negative growth rate* in total net cultivated areas has already set in suggesting no scope for horizontal expansion under food crops. Other moribundities are also there. A general plateauing of yield levels of major food crops with yield improvement of 0.1 per cent and I per cent in rice and wheat respectively during the last one decade is a matter of great concern.

It may howeve, be realized that *green revolution has not touched many areas and crops.* Even high fertile irrigated areas are still to experience such revolution. Instead of yield improvement, the declining output-input ratio has made farming less and less remunerative questioning the sustainability of our agriculture production and productivity systems. The increased concern for environmental safety, the degrading soil health, the increasing cases of pest resurgence and resistance and the exploding population are problems which question the ecological sustainability of our country. Last but not the least, with the increasing standard of living and per capita income, there will be increased demands for vegetables, fruits, pulses, oilseeds and value-added products.

Land Reforms

The fragmentation of land over the years impoverished the farmers. It has been found that out of 100 farming families, *75 per cent are marginal farmers with per capita holding of one hectare or less.*

With productivity tending towards constant return, as we have seen, the maladministration of land among cultivating households is bound to end up in greater *inequalities of income* unless remedied through appropriate reforms.[3] Land reforms towards preventing fragmentation of land *in particular,* have to be introduced.

One of the important aspects of land reforms in India is the creation of economic holdings. To establish economic holdings, the fixation of ceiling is required to be done so that those who have more than the economic holdings will have to surrender their surplus land to the public authorities who will then distribute the same among those who have uneconomic holdings. Another measure that can be adopted is to induce those farmers who have extremely small holdings to give up their lands and *shift to other occupations* in which case establishment of rural industries is required.

It is difficult, however, to create economic holdings. But a start should be made with the *laws of succession*. The present system of inheritance according to the Hindu and Mohammedan laws of succession, allows all sons and daughters to share in the ancestral landed property. This system should be so changed that partitioning of land below the size of economic holdings is made illegal.

Post-Harvest Losses

Post-harvest handling to prevent processing, transport and storage losses is very important. It has been estimated that if one half of these losses amounting to the tune of 20-25 per cent of the total food grain production, could be reduced, it will be sufficient to meet the projected food requirements till the year 2007. An efficient product handling system, proper transport chain and information network are essential to prevent the huge losses. A better storage system can further reduce the colossal waste of foodgrains that has been reported yearly.

Rural Infrastructure

The entire rural infrastructure has to be refurbished to make rural activities viable. The public sector investment in rural areas which has been declining now, will have to be stepped up especially in building rural transport facilities and minor irrigation works. These will help exploitation of waste and fallow lands accounting for more than 60 per cent of the cultivable area. Productivity in India is very low compared to world averages. Though this has been attributed to the climatic advantage that temperate countries enjoy but within the tropical belt too, India is far behind Brazil, Japan, etc. Within the country itself there are large productivity gaps between the higher productivity and lower productivity areas. This anomaly is seen even within the irrigated areas also. In rice alone, it has been found-by the ICAR that about *32 per cent of the irrigated areas* is of low productivity.[4] Thus there are great potentials to step up the yield level of foodgrains. Greater efforts will have to be made at raising productivity in the less fertile regions through better irrigation facilities, access to cheap farm inputs and storage facilities.

Rural-urban Migration

As the agricultural sector matures attaining higher levels of land and labour productivity, the demand for more specialized agro-processing, more sophisticated trading and repair services and more professional education, health and personal services would increase. All of these *require investment* and would tend to raise the share of non-agricultural employment especially *due to the push of labour out of agriculture associated* with mechanization. A degree of occupational *diversification* that offers secondary or alternative employment outside crop production would take place reducing the squeeze on standards of living that excessive population pressure on land implies. It would also mitigate the worst effects of such a squeeze such as the excessive rural-urban migration that has been witnessed by the slums and shanty towns in different monopolises. A prosperous agriculture is therefore the only determining factor for jobs for the rural poor.

The foregoing analysis of the reasons responsible for the failure on the food front for the poor, has also brought to light that the challenges that lie ahead are formidable. But their solution may be gratifying. We would deal in the succeeding paragraphs only two aspects of the remedial measures that could be taken to guarantee food security to the poor sections of society now and in the future.

Poverty Alleviation

The properly designed poverty alleviation programmes have to be evolved so as to guarantee food security to the rural poor. This could be done by increasing *the purchasing power* of the rural poor.

While effective programmes for both rural and urban poor are necessary, wage and self-employment in the rural areas are both a must to reduce the migratory flow of the rural population to the cities thereby prevents the increased urban poverty.

It is imperative to extend planning, infrastructure, appropriate technology and complimentary resources and services to the rural areas as a matter of priority.[5] The hurdles which have stood in the way of such development in the past seem to have diminished to a considerable extent in recent years and appropriate strategies for improving the rural-urban balance need to be devised immediately. Rural areas could then hope to enjoy some of the advantages of big cities like electrification, better communication system, minimum needs, sufficient purchasing power and facilities to live a normal contented life. And, the process of migration to the urban centres could then be checked to a certain extent and ensure rural-urban balance.

Public Distribution System and Nutrition

The operation of the government-run Public Distribution System (PDS) during the *past four decades*, shows that while it has withstood the test of time as an *effective food delivery system* it still needs further refinements to make it relevant to the emerging new economic order which necessitates reduction in overall food subsidy without jeopardising the interests of the poor. The PDS as at present, has been criticised as being urban-biased. A study based on the National Sample Survey's 42nd Round (NSSO, 1990) has revealed that while the overall subsidy per family is almost Rs.12.50 per month, it is hardly Rs.2.08 *for the bottom 20 per cent* of the rural population.[6]

Attempts must be made, at restructuring and better targetting the PDS to make a substantial difference and to ensure that more food can reach the poor. *The concept of security must* also change. The stress is to be more *on nutrition security* than mere provision of food. In fact, a national nutrition policy was adopted in 1993 followed by the initiation of the national plan of action in 1995. The Ninth Plan Approach Paper has taken a broader view of food security to include nutrition security as well. It categorically states that food and nutrition security is one of the primary objectives of the Ninth Plan. It also calls for long-term measures for eradication of hunger and starvation by ensuring "availability, accessibility and affordability of balanced food and nutrition for all". To ensure a more varied nutritionally balanced diet, farm incomes are also to be enhanced by other agro-based commercial activities that have been dealt with above.[7]

North-East India

The cut in subsidy on PDS and its complete withdrawal on APL (Above Poverty Line) may affect the poor in the North-East Region more than those in other parts of the country. As far as BPL (Below Poverty Line) is concerned, the amounts of foodgrains have been doubled but prices have been increased. Thus the poor would in no way be freed from the burden despite the assurance of the Central Government not to create any such burden. This is especially true with the poor people living in the remote and backward areas of the region.

With the unprecedented success in rice production and its average yield per hectare in Assam, the state is poised with the possibility to generate surplus foodgrains. The one lakh shallow tube well scheme under the Samariddha Krishak Yojana (SKY) implemented by the State has almost revolutionized the production of rice expected to break all past records with the farmers facing the problem of lack of space for storage.

The State agricultural marketing board should step in to fill the gap by procuring the surplus paddy on the minimum support price for the product. It should also set up processing units and take step to export rice to other North Eastern states where there is a shortage. This would be in line with the Centre's decision asking the states to procure foodgrains either from their own areas or from their neighbouring states to minimise transportation cost and transport loss of foodgrains.

Conclusion

The crucial problem is, however, the increasing demand for foodgrains which can be met only with a slower rate of population growth. There is the dire need to control the burgeoning numbers. As Swaminathan said, it is however becoming increasingly clear that without concerted national and global action to arrest environmental degradation and to check population growth, the Malthusian spectre may still be haunting the very first century of the new millennium.[8]

NOTES

1. United Nations Population Fund (UNFPA), Report 1999.
2. Report of the Indian Council of Agricultural Research (ICAR) 1999.
3. Saini, G.R., *Farm Size, Resource-Use Efficiency and Income Distribution* (1979), p.155.
4. I.C.A.R., op., cit.
5. U.N.F.P.A., op., cit.
6. "Ensuring Food Security" in *Yojana* (Vol. 41, No.1), January 1997. Also *N.S.S.'s 42nd Round* (NSSO, 1990).
7. The Ninth Plan Approach Paper.
8. M.S. Swaminathan, "Indian Agriculture: Looking Back and Forward: in *Yojana* (Vol.41: No.1, January, 1997), Publications Division, New Delhi.

5

FOOD SECURITY OF NORTH EAST INDIA FOR TWENTY FIRST CENTURY

— *S.N. Goswami, B.K. Sarma* and *S.B. Singh*

Food Security of North East India for Twenty First Century

The North Eastern Region of India falls under high rainfall zone and the climate ranges from sub-tropical to alpine. The region is characterised by difficult terrain, wide variation in slopes and altitudes, land tenure systems and cultivation practices. The agriculture of the region remains backward with low productivity. The region is deficient in food production and most of the food items are to be imported from outside. Therefore, there is an urgent need to increase the food production to make the region self-sufficient. This paper discusses the issues related to the above aspects, assess the production, requirement and deficit of foodgrains in the next 25 years and research efforts that should be adopted in future to make the North Eastern Region self sufficient in foodgrains production.

It is pertinent to discuss the present trend of production and productivity of foodgrains in the region. In this context the compound growth rate and contribution of area and productivity in foodgrains production are discussed here along with the foodgrains requirement in the year 2025-26 and future research perspective to achieve this goal.

Compound Growth Rate of Foodgrains (1980-81 to 1994-95)

The annual compound growth rate (per cent) of foodgrains in North Eastern States between 1980-81 and 1994-95 is shown in Table 5.1. It could be seen from the table that the annual compound growth rate of area in all the North Eastern States except Manipur, Sikkim and Tripura are positive signifying increase in area in these states.

Negative growth rate of area in case of Manipur, Sikkim and Tripura indicates decrease in area under foodgrains in those states. North Eastern Region as a whole, there was 9.69 per cent growth in the area which indicates overall increase in area under foodgrains.

Table 5.1 : Annual Compound Growth Rate (per cent) of foodgrains in North Eastern States (1980–81to 1994–95)

State	*Area*	*Foodgrains production*	*Yield*
Arunachal Pradesh	3.30	4.28	0.94
Assam	0.64	2.07	1.42
Manipur	(-) 1.54	1.53	3.14
Meghalaya	0.44	(-) 0.03	(-) 0.47
Mizoram	6.61	10.15	3.32
Nagaland	3.05	6.15	3.01
Sikkim	(-) 0.66	0.39	1.06
Tripura	(-) 1.02	0.95	2.02
N.E.Region	0.69	2.10	1.49

Annual compound growth rate of yield in all the North Eastern States except Meghalaya were positive indicating increase in productivity. In case of Meghalaya, the 0.47 per cent negative growth rate signifies fall in productivity of foodgrains. The highest growth rate of 3.32 per cent was noticed in case of Mizoram followed by 3.14 per cent in Manipur. A compound growth rate of 1.49 per cent was found for the North Eastern Region indicating an increase in productivity.

Overall productivity of foodgrains in all the states except Meghalaya was increasing. This was indicated by the positive growth rate in those states. However, the negative 0.03 per cent growth rate in Meghalaya indicates fall in the production of foodgrains.

Though there was negative growth rate in area in case of Manipur, Sikkim and Tripura, growth rate in production of foodgrains in those states was positive due to high positive growth rate in yield. This signifies that though there is decrease in area under foodgrains, the production increases in Manipur, Nagaland and Sikkim due to high growth rate in yield.

Contribution of Area and Productivity in Foodgrains Production

Table 5.2 shows the contribution of area and productivity in foodgrains production (pre cent) in North Eastern States. It could be seen from the table that both area and production contributed positively for foodgrains production in the States of Arunachal Pradesh, Assam, Mizoram, Nagaland, Tripura and North Eastern Region as a whole. In spite of negative contribution of area, productivity made a significantly high contribution in increasing the overall production of foodgrains in Manipur and Sikkim. In Meghalaya, though contribution from productivity was negative contribution of area in increasing production was significantly high. For the North Eastern Region as a whole the contribution of area and productivity to the foodgrains production was to the tune of 28.7 and 71.3 per cent.

Table 5.2 : Contribution of area and productivity in foodgrains production (%) in N.E. States (1980-81 to 1994 - 95)

States	*Area*	*Productivity*
Arunachal Pradesh	73.0	27.0
Assam	28.5	71.5
Manipur	(-) 45.1	145.1
Meghalaya	192.7	(-)92.7
Mizoram	51.6	48.4
Nagaland	40.7	59.3
Sikkim	(-) 61.5	161.5
Tripura	49.9	51.1
N.E. Region	28.7	71.3

Foodgrains Production, Requirement and Deficit/Surplus

Foodgrains production, requirement and deficit/surplus in North Eastern States during 2000-01 and 2025-26 is shown in Table 5.3. It is revealed from the table that all the States of North Eastern Region except Arunachal Pradesh will be deficient in foodgrains requirement in 2000-01. Arunachal Pradesh will be surplus by 46,000 tonnes. Entire North East will be in deficit to the tune of 1997 thousand tonnes as far as foodgrains are concerned during 2000-01. The table also showed that the situation as regards foodgrains requirement is concerned would not be different during 2025-26. Only Arunachal Pradesh and Mizoram could attain surplus in foodgrains production. North Eastern Region as a whole the foodgrains deficit would increase from 19,97,000 tonnes in 2000-01 to 45,78,000 tonnes in 2025-26.

Table 5.3 : Foodgrains production, requirement and deficit/surplus in NE States (000' tonnes).

States	*2001-01*			*2025-26*		
	Prod.	*Req*	*Surplus/ deficit*	*Prod*	*Req.*	*Surplus/ deficit*
Arunachal Pradesh	274	228	46	543	503	40
Assam	4080	5448	–1368	6809	9363	–2554
Manipur	397	463	–66	583	879	–296
Meghalaya	162	458	–296	203	932	–629
Mizoram	171	186	–15	580	430	150
Nagaland	283	361	–78	608	1098	–490
Sikkim	102	108	–6	128	154	–26
Tripura	505	719	–214	733	1506	–773
Total	5974	7971	–1997	10287	14865	–4578

Research Perspectives for 21st Century

To bridge the gap between the production and requirement of foodgrains in North Eastern Region, research efforts require a reorientation. Research efforts on the following area have to be reformulated to enhance productivity of foodgrains:

Breeding crop plants

In order to make the region self sufficient in food, we need to double the productivity of all the food crops. Self-sufficiency in rice in the region is feasible by 2005. The region may also specialise in *Japonica* rice for Japanese market. Our cost of production will be low and thus, will have an advantage over other countries and also, within India. If the same pace of growth continues, the region will become an exporter of rice, rajmash, groundnut and soybean. In order to attain the above targets, the following measures need to be taken.

- — Development of blast resistant, phosphorus deficiency tolerant upland rice variety suitable for inters cropping with a productivity of 3-3.5 t/ha.
- — Development of blast resistant, iron toxicity tolerant waxy lowland rice varieties with 6-8 per panicles, 200-250 grains/ panicle, 25-30 g/1000 grain wt., 130-140 days duration possessing cold tolerance with harvest index 0.5 and yield 6-8 t/ha.

— Integration of biotechnology in rice breeding.
— Initiation of breeding programme in target environment for higher yielding groundnut, soybean and sesame. Intensive screening of pegion pea, rajmash, greengram and blackgram.
— Development of maize varieties with a yield potential of 6-7 t/ha with emphasis on popcorn varieties.
— Acceleration of breeding research for improvement of rainfed crops for higher yield, adaptability to varied topography and farming systems with special emphasis on lowland rice, maize, pulses and oilseeds.

Crop production

Agricultural productivity of the region and economy of the farmers can be achieved by adopting modern technology and ideal land use system. Double cropping of rice and other crops is possible in low valley land with adequate minor irrigation development. Unproductive upland rice could be substituted by high yielding potential crops like groundnut, soybean, sunflower, pegion pea, ete. Viable, sustainable and socially acceptable natural/organic farming could be done. On farm research and demonstration of improved package of practices using identified HYV of crops, recommended dose of fertilizers and use of plant protection measures should be emphasised. Identification of crops and their management under moisture stress and low temperature conditions along with organic/natural farming need to be stressed. Development of economical and ecofriendly crop production packages involving integrated nutrient, water and pest management technology is essential.

Crop protection

Use of chemical pesticides should be minimised to save the environment and health hazards keeping in view the loss of crops due to disease and pests. Since the region cannot afford loss of foodgrains due to pests, varieties with resistance/tolerance against major diseases and insect pests will have to be developed. Research on biological control of disease, insect pests and IPM strategy will have to be intensified.

6

AGRICULTURE SHOULD GET HIGH PRIORITY IN THE NORTH-EAST

— V.S. Mahajan

My mind goes back to the year 1959 when hordes of rats had descended in the territory of Mizoram and within a short time had destroyed everything including food and a large number of people had perished as a result of this forced famine. It was a neglect on the parts of government to rush supplies in time as well as exploitation by non-Mizo traders of this situation for their personal gain that had led to the emergence of Mizo National Front led by Laldenga which bought one of the bloodiest insurgencies for over two decades causing heavy losses to the Indian army when a large number of Jawans and officers were killed. Ultimately it was the 1986 accord signed between Mizo Front and Indian government that peace was restored and Mizo guerrillas who operated from thick jungles and other hide-outs came in the open. Many youngsters who had disappeared in jungles had grown into middle-aged men by this time and among them was the present Chief Minister of Mizoram Zoranthanga, a right hand man of Laldenga and had played a leading role in this long insurgency.

The upshot of the above is that this area of the North-East which is yet mostly sparsely populated, particularly interior parts comprising of Arunachal Pradesh, Nagaland and Mizoram, is highly vulnerable area and depends heavily on supplies of food and other essential commodities from outside. Further, because of difficult terrain where it is difficult to construct modern modes of communication, the situation becomes all the more vulnerable unless sufficient measures are adopted in time to stock essential commodities in good quantity.

In other words, in these areas we would have to move even beyond the conventional distribution system and would have tackle their needs on a priority basis. Unfortunately, despite more than 50 years of freedom this country has yet not learnt much in this respect and the usual complaint of the North-East is that even their genuine needs are not adequately met. Very recently Mr. Zoranthanga had very succinetly pointed at this and had added that Indian leaders including the present Vajpayee government would hardly think about this area while they were much involved in other areas. Thus there was need to devote more attention here and our leaders should also visit this area which would bring better understanding between the tribals here and the rest of the country.

In fact here we have the intelligentsia among tribals and also in states like Mizoram, Nagaland and Meghalaya the rate of literacy is higher than even several other areas in the country. Thus to keep these people in good humour and away from insurgency where we have had a bitter experience, we should give special attention to this far distant area.

We must encourage special measures to promote agriculture on a sustained basis using modern technology and inputs which alone would make it less dependent on food from outside. While settled agriculture in these inhospitable jungle lands and hill areas is a difficult task but then with determined effort and liberal investment it is not an impossible job.

In fact I recall in the mid-1980s under the leadership of Mr. Lalthanhawla, the then chief minister of Mizoram, a determined wave had arisen to change the topography of this area and in place of Jhum cultivation - which then accounted for over 80 per cent agriculture and was very injurious for local environment, converting a vast green land into virtual desert land, there should be settled cultivation and it was also pledged to complete this task before the end of the last century so that when we entered into the present millennium (which we have already done) Mizoram had practically zero Jhum cultivation.

I was also actively involved in this exercise and had prepared a 15-year plan, split into 3-sub plans for the government when gradually this process would be accomplished with sufficient support from the Planning Commission. While this had evoked considerable interest even from other governments where Jhum cultivation was still quite popular, I understand not much could be achieved for several reasons particularly lack of sufficient resources, resistance from farmers deeply entrenched in this .system from time immemorial as well as indifferent attitude of politicians and administrators.

Whatever might be the reason the fact remains that without determined effort it would be difficult to bring them the benefits of settled cultivation which alone would yield regular cultivation, higher production and productivity as well as save the local environment from further destruction. Above all it would make this area less vulnerable to food scarcity and bring steadily rising income to the farmers.

In the case of Assam I have often wondered that while this area is blessed with a large agricultural tract why it has also failed to make its proper use and continues to be dependent on imports. Much in this respect could be learnt from other parts of the country particularly from Punjab and Haryana who now raise 3 crops a year from the land which not long ago was difficult to cultivate. But then mechanization, modern inputs and determined farmers have changed the face of these areas which are not only self-sufficient but also export maximum quantity of surpluses as well as are the principal partners in building national food stocks. All this had also helped the farmers to earn maximum income where the average income is higher than those living in nearby urban areas.

Thus Assam and other areas where agricultural lands are available should follow the example of their North-West counterparts who have played such an important role in making India self-sufficient in food and thus saved it from costly imports. With liberal government assistance and with the cooperation of farm scientists and other experts, the situation can certainly be changed.

While all these steps are needed in the long-term development of this area when it has to think in terms of self-sufficiency, right now efforts need to be made to provide this area with sufficient succor through streamlining the public distribution system particularly in remote areas where adequate stocks of quality grains and other essential commodities should be built in time to face any possible eventuality.

Thus the Centre should ensure that per capita availability of grains and other essential commodities is quite sufficient to meet the needs of these 7 states and they no longer face vulnerable situation as they have been often facing

Also it would be desirable that we open other channels of trade with neighbouring countries like Myanmar and Bangladesh, even others for import of grains which would save us the cost of transportation involved in bringing food from distant parts of this country like Punjab and Haryana. Thus these trade channels would decidedly prove more beneficial.

In fact while we should tap all possible sources to build sufficient stocks of grain at the same time we should ensure that this exercise is done with great care. Otherwise we would be wasting precious public resources.

In fact right at the moment our own country is facing excess stocks of grains for which there is not sufficient space and a large quantity has been destroyed by keeping excess stocks in the open. Thus it is appropriate time for the North-East to build a good stock position from the excess grains already available within the country. This would save us running to neighbouring countries and save resources.

Thus we should make the best of existing opportunity to replenish stocks of food in the North East for quite sometime to come.

7

DETERMINANTS OF PRODUCTION AND INSTABILITY OF RICE IN NORTH EASTERN STATES OF INDIA

— *S.B. Singh, K.K. Datta* and *Kh. B. Singh*

Introduction

The North-Eastern part of the country, comprising of seven States Viz., Arunachal Pradesh, Assam, Manipur, Meghalaya, Nagaland and Tripura, is a predominantly rice producing region. This region accounts for 7.81 per cent of total area under rice and shares 6.07 per cent of the total rice production in India. But average per hectare yield of rice in the region (1426 kg) is far below the national average (1879 kg). Although rice production in the region has made significant strides, particularly since the inception of green revolution, there has been wide fluctuations in the rice production over the years and also, in different states of the region. Thus, it is of vital necessity to analyse the production and productivity trend of rice in North-East India where rice occupies 89.46 percent of the total area under foodgrains and contributes 92.32 per cent of the total foodgrains production. In this paper attempt has been made not only to examine the growth and instability in rice production in North-East India but also to investigate the factors determining production and productivity of rice in the region.

Methodology

Time series data on area, production and productivity of rice for the period 1972-73 to 1997-98 for the seven North-Eastern States and also for all India level were collected from the various issues of Basic Statistics of North-East India, Fertilizer Statistics and Area and production of Principal Crops in India and Economic Survey of

India. Specific period has been selected as the reorganization of North-Eastern States were completed only in 1972, Annual compound growth rates of area, production and productivity were calculated by using the following log linear function (Dandekar, 1980).

$Y_1 = a \quad (1+r)^t$................................(1)

Where,

Y= the value for which growth rate is to be calculated

t = time in years

r = growth rate.

Taking log both sides of equation.................. (1)

$\log Y_t = \log A + t \log (1+r)$

Putting $\log Y_t = Y$, $\log A = a$ and $\log (1+r) = b$

$Y = a+bt$

$1 + r = exp^b$ Therefore, $r = (exp^b - 1) \times 100$

Coefficient of Variation (C.V.) was calculated to study the instability in rice production. The contribution due to area, yield and interaction between area and yield to the increase in rice production of North-Eastearn States and all India was also calculated by using the equation (Sharma, 1977).

$\Delta Q = A_0.\Delta Y + Y_0.\Delta A + \Delta A.\Delta Y$....................(2)m

where, A_0 and Y_0 are area and yield per hectare in base year and ΔQ ΔY and ΔA are changes in production, yield per hectare and area, respectively between base year and 't'th year. The three terms on the right hand side of the equation when divided by ΔQ provide estimates of the contributions of average yield, area and their interactions (average yield X area) to the increase in rice production.

Results and Discussion

Growth rates of area, production and yield

Table 7.1 presents the state wise growth rates of area, production and yield of rice in North-East India. It can be observed from the table that during the period 1972-73 to 1997-98, the area under rice in the region increased marginally at the rate of 0.73 per cent per annum while the per hectare yield increased significantly at the rate of 1.61

per cent per year. The area expansion as well as yield increase caused the output to grow at an annual rate of 2.40 per cent which was significant at one per cent level. However, the annual growth rate of production of the region was lesser than the all India average of 3.13 per cent per annum. It was due to higher growth rate of per hectare yield in all India level as compared to North-East India. The low productivity growth rate in North-East India was due to lack of technological breakthrough in most of the States of the region.

Table 7.1 : Compound growth rates of area, production and yield of rice in North-East India (1972-73 to 1997-98)

States	*Area*	*Production*	*Yield*
Arunachal Pradesh	3-26**	4.15**	1.12**
Assam	0.83**	2.35**	1.44**
Manipur	(-) 0.39*	1.77**	2.08**
Meghalaya	0.14 ns	(-) 0.24 ns	(-) 0.36 ns
Mizoram	(-) 0.44 ns	4.10'*	4.40"*
Nagaland	2.78**	4.04**	1.52**
Tripura	(-) 0.85**	2.22**	3.08**
North East India	0.73**	2.40**	1.61**
All India	0.53*	3.13**	2.58**

* Significant at 5 per cent level; ** Significant at 1 per cent level; ns=Non Significant.

Out of seven States in North-East, four States have shown positive trends and three States have show negative trends in growth of area under rice. But except Arunachal Pradesh and Nagaland, the growth rates of area in other States of the region were significantly low (less than 1 per cent). It was due to the various restrictions imposed by the respective State Governments to the practice of *jhum* cultivation since the late eighties. Rates of growth of production and yield was observed to be positive in all the States except Meghalaya. Mizoram State experienced the highest growth rate of rice production (4.10 per cent) and the least being Meghalaya State (- 0.24 per cent). The growth rate of per hectare yield was also highest and lowest in these two States.

Contribution of Area and Average Yield to Rice Production

Contributions of area, yield and their interaction to the increasing rice production for the period 1972 - 1973 to 1997 - 1998 are depicted in Table 7.2. It could be seen from the table that the contribution of area in increasing rice production was appreciable in the States of Arunachal Pradesh, Meghalaya and Nagaland. Average yield was the sole contributing factor in increasing rice production in the states of Manipur, Mizoram and Ttipura (80, 139 and 101 per cent respectively). In Mizoram and Tripura, the significant yield effect offset the negative area effect. Area and average yield contributed almost equally in increasing the rice production in Assam. The highest contribution of 36 per cent from the interaction component was noticed in Nagaland. It is interesting to note that except Mizoram the interaction effect was positive in all the States of North-East India. On an average, the contribution of average yield in increasing rice production in North-East India (60 per cent) was lower than the all India level (71 per cent).

Table 7.2 : Percentage Contribution of Area, Average Yield and their Interaction.

States	*Area*	*Yield*	*Interaction*
Arunachal Pradesh	63	21	16
Assam	42	48	10
Manipur	9	80	11
Meghalaya	80	18	2
Mizoram	(-) 21	139	(-) 18
Nagaland	43	21	36
Tripura	(-)8	101	7
North East India	28	60	12
All India	16	71	13

Instability in Rice Production

Rice production in India has remained moderately been stable over the period 1972-73 to 1997-98 at a low level of productivity. The average production during the period was 60,098.2 thousand tonnes with a coefficient of variation (C.V.) of 23.42 per cent. Rice production in North-East India was more stable than all India level with an average of 3714.32 thousand tonnes and coefficient of variation (C.V.) of 18.59 per cent (Table 7.3).

In North-East India, the maximum variation in rice production was observed in Mizoram (C.V. 45.47 per cent) and the minimum in Meghalaya (C.V. 8.54 per cent). These are the two States which

Table 7.3 : Instability in Rice Production in North East India (1972-73 to 1997-98)

States	*Area (000' hectares)*		*Production (000' tonnes)*		*Yield (kg per hectare*	
	Average	*C.V.(%)*	*Average*	*C.V.(%)*	*Average*	*C.V.(%)*
Arunachal Pradesh	97.13	23.74	103.23	28.98	1058.36	11.57
Assam	2831.88	6.58	2638.01	18.60	1114.12	12.73
Manipur	164.18	6.78	282.38	18.16	1716.85	19.63
Meghalaya	105.22	3.56	117.95	8.54	1125.20	8.52
Mizoram	57.57	26.19	56.39	45.47	996.20	37.49
Nagaland	105.36	26.47	107.37	44.49	1012.84	16.64
Tripura.	275.36	8.86	400.44	18.25	1458.88	22.37
North East India	3134.57	5.78	3714.32	18.59	1170.60	12.94
All India	40652.68	4.50	60098.2	23.42	1467.15	19.49

showed the highest and lowest growth rates of rice production in the region. The variation in rice area was much smaller than that in production both at all India and North-East India levels. The maximum variation in area was observed in Nagaland minimum in Meghalaya in North East India. The yield per hectare was stabilised at a very low level in all the States except the States of Manipur, Mizoram and Tripura where the coefficient of variation was as high as 19.63 per cent, 37.49 per cent and 22.37 per cent respectively, It revealed that the States with higher growth rate of yield per hectare witnessed high instability in rice production, Mehra (1981) Pal and Sirohi (1989) reported similar findings in their studies in instability in crop production in India in the context of new technology.

Reasons for Low Productivity

Table 7.4 presents the factors associated with growth of rice productivity in different states of the region. The table indicates that the States with higher productivity growth rates were having relatively lower size of holdings, For instances, the average size of holding in the leading growth states of Mizoram, Tripura and Manipur ranged from 0.97 to 1.38 hectares as against 1.77 to 6.82 hectares in the low growth States of Meghalaya, Arunachal Pradesh and Nagaland. The average size of all India level (1.57 ha) was also lower than the North-East average (2.44 ha). It shows high negative relationship between farm size and rice productivity.

The average percentage area under high yielding varieties for North-East India being 50.42 was far below the national average of 65.90 per cent. In the higher growth States of Mizoram, Tripura and Manipur, the percentage varied from 37.5 to 73.55. It reveals that area under high yielding varieties have direct bearing on the productivity of rice in the region.

Table 7.4 : Factors Affecting Growth of Rice Productivity in North-East India.

State	*Average% Size of holding (ha) (1990-91)*	*area under in HYV's of rice of net (1992-93) area (1975-1990)*	*%change in fertilizer proportion per cropped irrigated (1978-1990)*	*%change use in kg area (ha) (1000 ha)*	*Intensity of tractoris-ation (No. per (1990-91)*	*Use of Pesticid-es in kg. cropp ed area (1990-91)*	*Credit availa-bility (Rs ha) (1990-91)*
Arunachal Pradesh	3.62	22.61	5.58	334.78	0.94	0.10	419
Assam	1.31	51.88	(-)4.09	263.16	3.63	0.13	725
Manipur	1,23	46.67	0.00	253.90	3.27	0.20	605
Meghalaya	1.77	37.5	(-)7.69	32.26	1.79	0.18	1222
Mizoram	1.38	37.5	33.93	864.71	2.11	0.10	1305
Nagaland	6.82	24.44	(-)4.89	164.29	7.51	0.06	1251
Tripura	0.97	73.55	21.60	587.10	0.38	0.31	780
N.E.India	2.44	50.42	8.07	247.52	2.8	0.19	901
All India	1.57	65.90	37.04	142.42	8.72	0.37	1046

Mizoram and Tripura States which were the centres of high productivity growth had the highest percentage changes in proportion of net-irrigated area (33.93 and 21.60 per cent, respectively). Percentage change in fertilizer use per hectare was also high in Mizoram and Tripura States (864.71 and 587.10 per cent, respectively). It follows that fertilizer use had almost a complementary relationship with the availability of irrigation facilities and finally, increased the rice productivity. Similarly, the availability of credit was highest in Mizoram and the least being Arunachal Pradesh State.

As shown by the North-East average of 2.80 as compared to the national average of 8.72 tractors per 1000 hectares reveals that agriculture in the North-East India is highly labour intensive. In States such as Assam, Manipur, and Mizoram which have comparatively higher growth rates showed higher rates of tractorisation. The use of pesticides was also found to be on the higher side in the higher growth States of Tripura, Assam and Manipur.

Conclusion

From the above discussion it has been concluded that only Mizoram and Tripura States experienced higher growth rate of rice productivity. The remaining States witnessed growth rate lower than the national average. Increase in rice production in the region was mainly due to the improvement in yield rather than area expansion. The States with higher growth rate witnessed high instability in rice production. Among the various factors, availability of irrigation facilities, adoption of high yielding varieties (HYV) of rice, rate of fertiliser use, farm size and credit availability experienced great impact on rice productivity of the region. Thus, the major policy thrust in this context should be on enhancing irrigation facilities in the States which were lagged behind. Development of water harvesting techniques such as micro-watershed based farming system on hill areas can go a long way in increasing rice productivity in the region. There is enough scope to narrow down the adoption gap in the use of fertilizers and high yielding varieties in various States through intensification of extension services and institutional support. The existing pattern of tractorisation in the region was found to be not significant as compared to the other States of the country. On this context implementation of land revenue act and conversion of *Jhum* fields into terrace fields in hilly areas need to be emphasised. Besides, it will further scale down the average size of holding and consequently, will effect the agricultural productivity. Financing and credit policy of the financing institutions should be liberalised. These measures will facilitate on equitable growth of rice production among the States and overall development of the region.

REFERENCES

Dandekar, V.M. (1980). Introduction to Lonavla seminar on data base methodology for the study of growth rates in agriculture, Indian Journal of Agricultural Economics. 30 (2) : 6.

Mehra, S. (1981). Instability in Indian Agriculture in the Context of the New Technology, Research Report No. 25, International Food Policy Research Institute, Washington DC. U.S.A.

Pal, Suresh and Sirohi, A.S. (1989). Instability in Indian crop production: its magnitude and sources, *Artha Vijnana*. 31(3) : 241-256.

Sharma, K.L. (1977). Measurement of the area, yield and prices in the increase value of crop output in India, *Agricultural Situation in India*. 32 (96) : 349-51.

8

THE 21st CENTURY AND THE EMERGING SCENARIO AND ISSUES OF SOCIAL SCIENCE RESEARCH IN NORTH-EAST INDIA

— K. Alam

The North-Eastern part of India is a land of vast diversities and is stooped in its myriad problems. Its rich natural resources are yet to be fully exploited and hence the region continues to remain underdeveloped, The North-Eastern region as a whole is relatively underdeveloped compared with the other regions of the country. It is only after the independence that development process in the region began. Infrastructure facilities, so vital for economic development, are lacking in the region. Power shortage is immense in spite of the large hydro-electric power potential of the region; moreover, the region is not well connected with the rest of the country by roadways and railways; the coverage by institutional banking is poor; the region is not able to attract enough investment from outside to take it to a higher growth plane. More importantly, the human development scenario is worse. Lack of education and health-care facilities means that the skill-base of the populace is negligible and hence its productive potential is too small to allow a sizeable income growth to take place. In the agricultural sector, during the outgoing century, green revolution failed to take place in the region. Subsistence agriculture continues to the predominant system of agriculture in the region as before.

The Industrial development of the country began with the decisive shift in strategy via the 2nd Five Year Plan and the Industrial Policy Resolutions pronounced in 1948 and 1956. The public sector was given a prominent role in the entire process. However, the role of public sector in the industrial development of the region was marginal, to speak the truth, whatever industrial development took place during the last century was sporadic and of low level. Operation of some of these industries, the paper mills in the central sector and the plywood factories in the private sector resulted in overexploitation of natural resources, especially forest resources, which led to ecological imbalances in the region. Coupled with the fact that many a primitive method of earning livelihood by the indigenous population persisted added to the pulls and strains on our fragile ecology.

The above-mentioned issues were a few of the areas of social science research in the North-East India. But as we cannot expect the socio-economic perspective to undergo a fundamental change in the new century at least in the initial period, many of the socio-economic problems of the outgoing century would naturally spill over to the new century. Besides, some new issues are likely to emerge in the present century and would demand serious attention in the research programme of the social scientists. We intend to discuss a few of the more important issues.

Issues of Environment

The most important issue of social science research during the 21st century should be regarding the need of environmental preservation and ecological balance, which has been seriously disturbed by the ruthless exploitation of nature for economic development. But the need to solve the basic problems of the economy underdevelopment, poverty, unemployment, malnutrition, etc, cannot be solved without the use of natural resources. The reconciliation of both development and conservation leads to the new concept of development strategy called sustainable development.

The role of natural resources in the economic development of a country is paramount. The logic is simple- economically exploitable natural resources add to the productive capacity of an economy; greater the resource availability, greater the number of times it can turn a machine and so greater is the output in the economy. The North-East has abundant mineral resources, mainly in the form of coal and limestone, oil and natural gas, which are being extracted in commercial quantities from the oilfields in Upper Assam. Forests play an important role in the economy of the region. The proportion of forest area to the total area is quite substantial. But unfortunately during the last few decades, the area of forests as per cent of total area has fallen due to large-scale deforestation and encroachment.

Assam and Arunachal Pradesh are also rich in water resources. There are mainly two categories of water resources- underground water and surface water. In Assam, there is plenty of surface water because of her wide network of river system. Assam has two river systems- the Brahmaputra and its tributaries in the north and the Barak-Kushiara river system and their tributaries in the South. The principal uses to which these vast water resources can be put are irrigation, hydro-electricity generation, navigation, fish culture etc. The Central Water and Power Commission made a survey of power potentials of these rivers during 1953 to 1958. The Survey revealed that the large dependable discharges of the Brahmaputra and its two major tributaries, the Lohit and the Dehang contributed by their snow covered catchment areas and their steep course through the Himalayas constitute some of the major possibilities of river harnessing for economic development.

But the experience of the last few decades suggests that certain natural resources have been over-exploited, while on the other hand, others remain under-exploited. Over-exploitation of forest products by both forest dwellers, who cut down trees to earn their livelihood, and by plywood industrialists, for use as raw material, is a matter of grave concern. But along with the concern for maintaining ecological balance, due consideration must also be laid on the following —

First, the villagers dwelling in the periphery of the forests have a hand to mouth existence. Their resource base is negligible. The forest produce they extract is all they earn as their livelihood. If in the name of environmental protection, these lowest rung of our society are deprived of their only source of living, then the development process will not only bypass them but will also be meaningless.

Secondly, the forest resources of North-East India provide the raw material for several industries, especially plywood industries. Plywood industries are labour intensive and hence can help ease the grave unemployment situation of the region. Besides the mineral resources, the forest resources have helped in the setting up of a few large industries. This has facilitated the industrial development of the region. But the industrial development has been mostly unsustainable on account of its lack of concern for the environment and ecology. Over-exploitation means that extraction of resources from their natural environment is much more than its replenishment, thereby living a diminishing quantity of it for our posterity. Maintenance of interm generational equity is thus not only an ethical question but also a question of survival of our future. But considering the two above-mentioned perspectives, there must be a drive towards achieving even intra-generational and inter-regional equity. Here comes the concept of sustainable development. Intrinsic to this concept is the act of balancing development and preservation needs, policy measures for which must be framed after an intensive study of all its aspects, which must be the job of our social scientists.

Issues of Economic Development

Economic development has eluded the North-Eastern region for the last fifty years of plan development measures. In the near future also there is very little hope of speeding up the development process because of certain deterrent forces working in the region. The problem can be briefly analysed as thus:

North-Eastern India is one of the industrially most backward regions of the country. Industrial investment in the region is one of the lowest in India. This is mainly because of three reasons. One, the income level of the people is very low, and two, the region is able to attract very negligible investment from outside. Three, local entrepreneurship is conspicuous by its absence.

The low-income level is primarily due to lower productivity in agriculture where subsistence farming continues to exist and the benefits of improved technology ushered in by the Green Revolution have scarcely reached the farmers of the region. Also lower investment results in lower income level of the people.

Investment from outside the North-East is very small on account of a host of factors. The most serious one is the problem of insurgency and its associated problems of violence, abduction, extortion, etc. creating a fear psychosis in the minds of investors. Economic development has been the greatest casualty of this abnormal situation. North-East India is a multi-ethnic region. The recent years have witnessed aggravating unrest among the major tribes of the North-East as a result of persistent economic backwardness and an intense feeling of deprivation and neglect. The unemployment scenario is worse. Unemployed youths, out of sheer frustration and a feeling of hopelessness, easily get lured to a life outside of the mainstream of society. The linguistic divide and the tribe-non-tribe hatred have led to an increase in violence against the people from outside. Hence the region is able to attract very little investment from outside. Social scientists should accord high priority to this area and an in-depth study should be made on the problem in the coming years.

Lower investment is also on account of very poor infrastructure facilities. Inspite of having the largest hydroelectric power potential in the country, the actual power generated is far short of the demand. Hence power shortage disrupts production in the industries of the region. Inadequacy of road and rail facilities increases the transportation time and cost. Poor coverage by institutional banking means that the credit flow, so vital for augmenting industrial investment in the region, is low.

Lower investment coupled with lower productive efficiency, which is denoted by *Incremental Capital Output Ratio,* has led to the poor economic growth as has been experienced till today. Lower productive efficiency is primarily the result of two, factors one, poor infrastructure (which has already been discussed earlier); two, lower efficiency with which the work population turns out the output and this is due to the low knowledge and skill-base, a reflection on the state of the precarious education and health facilities in the region.

Thus what we observe from the analysis made above is the presence of a *vicious circle of underdevelopment* in North-East India. Low income leads to low industrial investment. Lower income of the people means that expenditure on health and education is affected; thus the skill-base of the population becomes low enough to depress the productive efficiency and lower economic growth. Lower economic growth worsens the unemployment situation in the region, a major cause of social unrest, manifested the nefarious militant activities. Such a fluid state of affairs hampers investment from outside, thereby further depressing economic growth. And the circle is complete. Thus social research must be directed towards breaking this *vicious circle* and ushering in a new and higher growth path.

Issues of Human Development

Another important issue to be taken up in the realm of social research in the 21st century is human development. It has now been realised by the Third World economists that economic growth is meaningless without human development. Human development and not merely economic growth should be the criteria of success of the development efforts of the society. Economic development is an all-inclusive concept, which apart from considering economic growth of an economy, takes into account the following conditions: poverty, unemployment and inequalities of income and wealth. Economic development should improve the quality of life through egalitarian measures and adequate provision of basic needs of life including education and health services, which is reflected in the Human Development Index (HDI) annually prepared by the UNDP.

The poverty scenario in the region is very upsetting. First of all, there is poor economic growth. And then whatever growth has been achieved; the fruits have mostly been appropriated by the upper strata. This has increased income and wealth inequalities. The consequences are there for everyone to see. The sharp rich-poor divide has increased anti-social activities; the crime rate has increased enormously, and the hatred for the people from outside, whom some of the ethnic populace considers as being responsible for the poverty of the large section of the people. High poverty prevalent among the population is on account of low income, which in turn is due to two factors: one, people have very low income-earning assets, because the income they earn is not adequate enough to invest in a high-yielding assets, and two, the knowledge and skill-base of the population is low because the income level is again not adequate enough to incur expenditure on education and professional training, so that they could offer only low value-added services which will invariably provide them low returns. Thus in both the cases, what we observe is the operation of a *vicious circle of poverty*, which has to be broken by suitable policy measures to be designed after the social scientists makes an indepth study.

A very important consequence of poor economic growth has been the hopeless employment situation prevailing in the North-Eastern region. A minimum of capital investment is required to generate meaningful employment. Investment being low in the region on account of a host of factors discussed earlier, the scope for generating gainful employment is limited. Lack of entrepreneurship and lack of availability of institutional credit has been the main reasons why self-employment is not a serious career option of an average youth today. What we observe in agriculture is the presence of disguised unemployment due to too much population pressure on land and the absence of an alternative employment in the non-farm sector. The consequences of unemployment, as seen today, are mainly two: acute poverty and the consequent feeling of neglect among the masses, and the increasing frustration among the youth manifested in the rise of militant activities. Social research in the 21st century must face the great challenge of solving this vexed problem of unemployment.

Socio-cultural Issues

The North-East is home to numerous races and tribes. People have lived in relative peace and unity through the centuries. But of late, racial acrimony has surfaced due to certain factors. Ethnic consciousness has taken roots among the diverse groups generating a sense of separatism. People no longer think in terms of a greater brotherhood. They now want to identify themselves in terms of their own races and tribes. In asserting their independent identity, they often come into conflict with one another and with the mainstream society in general. This has led to the increasing tension in our society.

Then there is the hatred for people originally belonging to regions outside the North-East. These people have made North-East their home long ago and have contributed significantly towards the economic development of the region. But a mindset that has come to dominate the local psychology is that these people have exploited the local population and taking away more than their justified fruits of labour. They are suspected to be the main reason behind their poor economic condition. Such a mindset and a bitter acrimony are dangerous for the peaceful co-existence of people in the region. Therefore, it behoves on the social scientists to make a serious study on their socio-economic conditions and also on their socio-cultural traits, which have an impact on their economy in the new century. Moreover, large-scale influx of people from Bangladesh has created a host of social and economic problems for the region. The local people consider this relentless influx of migrants as some sort of silent invasion of their land. The problem has been confounded by the fact that many of these immigrant people came and settles down permanently long back and they have been completely mixed up with the local people. Therefore, it has become extremely difficult now on the part of the Government to identify the illegal migrants in order to deport them. This has created social, economic and political problems for the region, which defy solution in spite of persistent popular upsurge and Governmental efforts. The social scientists have an important role to play in making an in-depth study of this complex problem in all its aspects in the new century.

There are certain other aspects as well, which will demand the attention of social scientists, especially the inter-mingling of outside culture with the indigenous tribal culture. Preservation of tribal culture in its entirety may not be possible on account of the need to contact the outside world. But negative influence on it can be avoided if proper study is made on the nature of contact, the extent of influence and so on.

Gender Issues

Women studies and economic development should be given due priority in the research programme of the 21st century in view of the importance of the role of women in economic activities which was long ignored. Women empowerment, women education, gainful employment of women through entrepreneurial programmes and self-employment schemes, role of media in highlighting women issues, professional attainments of women of the North-East, comparative study of the position of women in the North-East vis-a-vis the social position of women in the rest of India, etc. may be some of the areas of interest for the social scientists in the North-Eastern region. Women have long carried the burden of the family but the returns to them in the form of provision for education, health and welfare facilities have not been commensurate to the contribution they make towards their families. To promote gender equality, various aspects of gender discrimination must be studied in detail and sensitised among the masses. The matrilineal societies of the North-East can form some sort of a backdrop to base such studies.

Issues of Land and Agriculture

The whole of the North-Eastern region's economy is mainly based on agriculture. It is not only the economic condition of the people, but also their social and cultural characteristics that are largely shaped by agriculture. But the basic problems of agriculture of the region have not been properly addressed during the outgoing century. The crucial problems of agriculture that plague the region for a long time are loss of land or land alienation, both inter-community and inter-class, dwindling size of the landholdings affecting farm productivity which has aggravated rural poverty and income inequalities, transformation pattern of shifting cultivation in the hill areas, lag in the implementation of land reforms, problems of agricultural finance, marketing and other services etc. All these problems deserve attention of the social scientists in their programme of research in the new century.

Other Issues

The other priority areas of social science research in the 21st century are Decentralised Planning and Panchayati Raj Institutions (PRIs), importance of border trade in most of the North-Eastern states problem of subsistence farming and commercialisation of agriculture, entrepreneurship and management capacity, motivation and behavioural patterns and their relationship with economic development, etc. More other priority areas of research are development strategy for North-East India, population studies with special reference to human resource formation and manpower planning, resource appraisal with emphasis on preparation of inventories resources and resource utilisation etc., inter-dependence between the North-Eastern region and the rest of India as well as units of the region.

9

SOME ISSUES OF SOCIAL SCIENCE RESEARCH IN NORTH-EAST INDIA IN THE 21st CENTURY

— *S.B. Chakrabarti*

Introduction

Legacy of inheritance in social science research in India has its obvious colonial hang-over. The formalization of the disciplines in social sciences in general became functionally accountable during the 20th Century. As a natural process of social development one could feel inclined, before trying to understand the emerging factors, to map out the by-gone social processes in the immediate past. If one can pin point on the more dominant driving social, economic, political and cultural factors one would be able to study its real impact on the life of the people in a given geographical location.

Social research is a special kind of involvement in quest of knowledge where causalities are attempted to be established on the recurrent and varied set of social phenomena. Once the relations between two or more sets of facts are identified, some kinds of predictions are possible to make towards the trend of changes across these facts. The question of priority is dependent on very many individual, institutional, social, political and economic criteria. It is not the question of convenience of handling a priority, but it is also to be worked out in terms of socio-economic and politico-historical cost benefit paradigm. Considering the great limitation of our resources - both human and financial — we must try to reduce the burden by minimizing duplication of work at the first instance and then making the result of our research immediately available for public consumption so that useful application of the same, if any, may be made in the larger interest of all concerned.

Logically it follows that we should have a proper documentation of what has already been done. Next we should identify the obvious gaps in information that need to be filled either by computing data from the available secondary sources or by generating fresh data at the primary level by way of undertaking field research at the appropriate context.

Objective

The present paper makes an attempt to understand some problems and the emerging issues in North-East India during the 21st Century. The broadly identified areas for this purpose would be following:

1. North-East India as conceived as a spatial terrain including its ecological peculiarities
2. North-East India as social aggregate and cultural entity
3. North-East India as economic zone
4. North-East India as religious enclave

North-East India as an unit of modernization and transformation

Spatial terrain

The North-East India as a politically conceived geographical patch of landmass has its wide ecological variety, climatic peculiarities, socio-economic diversities and so on. (Bhattacharjee, 1991)

While defining the boundaries of North-East Frontier of Bengal Mackenzie (1884) pointed out that "it embraces the whole of hill ranges North-East and South of the Assam valley as well as the Western slopes of the Great Mounting system lying between Bengal and independent Burma with its outlying spurs and ridges." He further mentioned, "They formed the Assam valley surrounded North-East and South by enormous savages and war like tribes whom the contacting authority of the Assam Dynasty has failed of late years to control and whom the disturbed condition of the province has insisted to encroachment". (p.7)

The Census of India in its publication mentioned that hill areas of North-East India mainly consist of Darjeeling district of West Bengal, autonomous hill district of Assam, North-East frontier Agency, Nagaland, the hilly tracts of Manipur and Tripura (Roy

Burman, 1961). A major part of the region of North-East included ancient Kamarupa and portion of the Neighbouring state of Bengal (Chaudhury, 1999). The North-East region has a total area of 255037 sq. kms. of which 4200 kms. covers international boundary with Bangladesh, Bhutan, China and Burma.

The North-East India as we know of today is somewhat an administrative creation after independence. A major part of today's North-East was primarily administered by the Britishers as part of the erstwhile state of Assam. While there was a patch known as North-East frontier Agency, which is present-day Arunachal Pradesh, since 1987 was left to special administration. The Nagaland (1963), Meghalaya (1972), Mizoram (1987) were later creations.

Today's seven sisters stand at different level of their socio-cultural integration in spite of having a somewhat common physiographic and socio-economic characters (Datta, 1992, Borah, 1994).

The North-East India is historically embedded in a cultural milieu belonging to south, south-east, east and central Asia. To conceive it as a microregion in isolation in the context of Indian sub-continent only could create some confusion understanding the region properly. It has to be looked at from the perception of wider contiguous area (Roy Burman, 1995).

Socio-cultural Entiry

In each of those states there are a number of communities who have their own social identities and cultural, religious, political and linguistic affiliations. After independence, there was on the one head a national effort of economic and administrative reconstruction and on the other a growing attitude on the part of some communities to spear-head the question of; identity along the narrow political line, sometimes even leading to violent methods. The climate of active dissent is still continuing. Therefore in all probability the 21st century would confront both these trends simultaneously, with of course the effort for sincere mediation by the national and regional political leaders. The emergence of middle class in the tribal heartland who arose from a tertiary sector often adopted ambitious line of personal agrandisement to the neglect of their poor fellow followers. There was still a continuity of nobility of the golden royal families in case of the princely states of Manipur and Tripura as well as the traditional Kings in case of Meghalaya.

In North-east India there has been an observable trend of vivesection of regional and state boundaries in Assam on the one hand to a continuous demand of enlarging a bigger area for particular ethnic group or groups on the other. For example, while Assam has been gradually reduced to give birth to some smaller states and just the opposite demand of greater Mizoram or greater Nagaland is also very active. Even, in a small state like Meghalaya, the ethnic bias is discernable in the political mobilization of various groups. For example, people often identify themselves either in favour of a Garo or a Khasi constellation in the over all political control system. In the political vicissitude, there has been a group dynamic in terms of a dominant-minority divide. Such syndrom has often been reflected in the normal democratic activities as well as in the insurgent activities in case of North-East. The introduction of the Autonomous District Councils which was basically meant for a self management of the local communities sometimes turned out to be a parallel administration when juxtaposed in the context of a state administration. The initial 'inner line' safeguards were sometimes not naturalized in tune with the spirit of the proposal. There is pressure both for preservation and extension of the same boundary maintenance mechanism as also the withdrawal of it in the greater interest of national development. The 21st century would in all probability confront all these existing forces mentioned above there has been a serious problem of influx of population from the neighbouring border areas and this is singularly an important factor which will cause great tension in case of North-east India in the 21st Century (Nandy, 1997). The apparent tribal non-tribal divide would still raise its hood either in defence or in offence in order to assert a cause or to defend a threat involving the local communities.

Economic Zone

The basic economic activity in most parts of North-East India except in the planes of Assam and the valley regions is dominated by the shifting cultivation. There is increasing pressure on land which resulted in the reduction of Jhum cycle. But the productive capacity of *Jhum* land has been unable to sustain the economic base of modern economic factors introduced through market and commercial net work (Ganguly, 1994). In the absence of developments of modern agriculture and sustainable industrialization which would absorb a sizeable section of rural and urban population, the North-East India has mainly depended on the other tertiary sectors like service, business, etc. which are obviously constrained by its own limiting factors.

The North-East India covers an area of 255037 sq. kms. According to 1981 census about 26.5 million people live in North-east India. They are 4 per cent of the total population of the country. The dominant economy of the region is shifting cultivation. It has both positive and negative effect on the land and people. Apart from the soil erosion, periodic floods, deforestation it also does a damage to valuable timbers. But on the other hand, given the hilly terrain, the technology of production could not develop first enough to accommodate sufficient people in agricultural production. Shifting cultivation also involves family labour and close-knit kin group to operate. Though the production is basically meant for consumption, lot of changes have also come about with the passage of time in terms of employment of technology, selection of crop and taking out some surplus to the market. In spite of the all out efforts, to reduce the area of shifting cultivation in North-East India and to replace it by alternative economic package, this has so far not been possible to achieve. Apart from the economic reasons social factors are also responsible for its survival. It is not going to be reduced even in the early 21st century.

In the 2lst century the North-East region is going to face severe problems of inelastic demand for agricultural product and improportionate labour utilization. In the absence of a developed agrarian set up and a dedicated peasantry the area will face the need of a regulated market and capital accumulation for further investment in productive activities. The slow but gradual increase of acreage in settled cultivation would virtually result in the growth of agrarian hierarchy, not in tune with the existing ethos of community ownership and management of land and its production. Even in some cases a terrific strain would emerge on the question of ownership and control of land along patriarchal or matriarchal line (Nongbri, 1988).

Religious enclave

The North-East India as a religious enclave presents broadly three types of religious faith, the Hinduism, Christianity and Buddhism of various segmentation. This apart, some local traditional movements are also appearing very strongly in certain parts of the region, eg. The Seng-Khasi in Meghalaya and Donyi-Poloism in Arunachal Pradesh. In each of these religious following sometimes differential political access has also been noticed from time to time depending on the currency of the movement of the local origin or blown up with the help of outside links. The tacit attempts of conversion also faced occasional resistance from within the local communities or being articulated by their supra-local network. With the forces gaining ground in terms of anti-conversion agitation on the one hand and a call for more liberalism in terms of accepting a particular faith on the other would probably confront each other in a specific situation.

Modernization and Transformation

In North-East India at one stage the modern developments were seen as inseparably linked with the developments of Christianity. For example, the power of the church organization in Nagaland in political affairs of that area was perceptible. The external network in case of the Buddhist cult also could be found in North-East region. Even those section who once identified themselves largely with the Hindu society have started claiming separate identity along the different religious line. In the recent years the political disturbances of many parts of North-East have often been associated with the rebellious activities based on different religious affiliations (Choudhury, 1994). There is possibility in the coming century of a face to face confrontation of religious fundamentalism on the one hand and a secular and liberal movement on the other. And in this another implicit issue may come on the surface, that is between the two major aggragation of population of tribes and non-tribes. The attitude of separatism and disruption might crop up at a very regular pace unlike the previous decades (Chaube, 1973).

The whole of this country in general and the North-East Indian states in particular are passing through some stages of transformation. While there are a common idiom of macro factors for all of them like the technological development, the monetized economy, the international network of trade and commerce, the overall bureaucratization of political and educational processes, the increasing trend for a higher rate of consumerism etc. North-East India is also known for having its own unique situations. For example, the relatively formidable land-man ratio, the apparent community based system of sharing the resources, the geographical isolation and its placement in the context of a number of international boundaries, the relatively lesser exposure to technological advancement, the constitutional safeguards for various security measures etc. have also contributed in their own way for the retention of some traditional social ethos on the one hand and to interact with the larger communities at per by way of participation in an uniform national, political, economic and administrative framework on the other. The regional peculiarities at work do no negate the plural societies and cultures of North-East India from undergoing the historical trend of newer social formations and its consequential transformation. Obviously, we are naturally confronted with the problem of growing ethnic consciousness and its manifestation in corporate community life on the one hand and the emerging problems of development at the cost of erosion of old moral values and rapid atomization of erosion of old moral values and rapid automization in social and cultural life on the other. These factors subject the people to a number of problems of adaptation and adjustment with their immediate environment, both physical and socio-cultural. A lot of conflicts and contradictions emanate from the formulation and implementation of the development projects. Research studies are very much necessary on each of these issues just mentioned. The entire period of the 21st century would probably face a number of alternative options. These will get effective, feed back provided we get to know the intricate mechanism of the controlling social, political, economic and cultural processes that keep the societies going from one level of consciousness to the other. In order to have first hand knowledge about such processes and systems (which is other wise known as the people's perception) one has to fall back on upon the rigorous research materials that would provide the solid intellectual base for tackling such problems. Al this stage this approach also should include the academic issues which have applied bearing.

In the North-East States the administration of the Sixth Schedule of self-management mechanism has been introduced by way of safeguarding the interest of the local communities mainly in saving its own resource base (land and forest) from being encroached upon by the powerful outside elements. The other aspects of it are to regulate a social and cultural autonomy on the basis of their own traditional mooring. But in the meanwhile slowly an erosion has started weakening the system mainly in the area of economic aspect where remarkable pace is set in privatizing land for self-interest rather than the community interest. Therefore, the 21st would definitely face a question as to how the land is going to the market even in such a constitutionally guarded self-management socio-economic set up. And there will be enough scope to study the problem of land, market forces and trade activities in the North-East India in the years to come. Another very interesting area of inquiry in the 21st century could be the contradiction of the socio-political and economic control in these autonomous institutions vis-a-vis the management of political bureaucratic system. There would probably be an enforcing pressure from the higher authoritative structure of the state on these autonomous institutions on the one hand and on the other a kind of elite manipulation of the power structure which help widening the gap between the masses and the managers of the systems. It would be interesting also to study how instead of all these monstrous supra-local and macro-dimensional forces remaining at work the North-East India by and large is able to keep its traditional community bases resource management system. There would be obviously a related question to ponder about in the near future as to how Micro-level social and political movements gather its momentum in the face of an overall regional economic imbalance. As a corollary to this the outflow and in-flow of the resources in a given area under study would also require an in-depth inquiry by the scholars.

Some Urgent Issues for Further Studies

1. Marginal communities and their adaptive strategy
2. Productive Organisation and the changing strategy
3. Pattern of urban transformation and the new community building
4. Ethnic mobilization and the emerging coalitions
5. Primordial Institutions in the changing phases
6. Religious movements and its twists and turns
7. Civil society vis-a-vis the adaptive mechanism of modernization
8. Rural urban migration and the new trends
9. Pattern of consumer society and the recent developments

10. The new generation and the problems of communication
11. Agricultural development and the new agrarian relations
12. Tribal arts and crafts and the dwindling nature
13. Population and occupation and their structural shifts
14. Class formation in a new political and bureaucratic hierarchy
15. Role of technology in social transformation
16. Recent developments on crime, drug, border trade, smuggling, video - piracy etc
17. Self - management institutions and their optimal utilization
18. Economic development and the question of insurgency

REFERENCES

Bhagabati, A.C. 1988. 'Tribal Transformation in North-East India : An Apprisal of Emerging Ideological Dimension' Presidential Address, Section of Anthropology and Archaeology, 75th Indian Science Congress, Pune

Bhattacharjee, J.B. 1991. *Social and Polity Formations in Pre- colonial North-East India*, New Delhi, Vikas Publishing House Pvt. Ltd.

Bhowmick, Subir , 'Patterns of Minority Violence in North-East' (mimeographed).

Borah, Dhrubajyoti 1994 'Understanding Assam and the North-East', *The Administrator*, Vol.XXXIX pp.63-76

Chaubey, S.K. 1973. *Hills Politics in North-East-India*, Calcutta : Orient Longman.

Choudhury, Rabijit 1994, 'Religious Undertone in Ethnic Issues in North-East "*The Administrator*, Vol.XXXIX pp. 15-31

Choudhury, Sujit 1999. The North-East : A Concept Re-examined', *Seminar*, No.474

Datta, P. S. 1992. *India's North-East : A Study in Transition*, New Delhi : Vikas Publishing House Pvt. Ltd.

Ganguly, J.B. 1994. 'The Economics of Transition from Shifting to Settled Cultivation in North-East India, *Journal of the Indian Anthropological Society*, Vol.29 No. 1 & 2, pp. 75-81

Mackenzie, Alexander 1894. History of the Relations of the Govt. with the Hill Tribes of the North- East Frontier of Bengal Calcutta : Home Department Press

Nandy, Sibdas 1997. *Infiltration in North-East India : Fact and Fiction*, The Window, pp. 21-34

Nongbri, Tiplut 1988. 'Gender and the Khasi Family Structure : Some Implications of the Meghalaya Succession to Self-acquired Property Act.1984', *Sociological Bulletin*, Vol.37, No. 1 & 2, pp.71-81

Roy Burman, B. K. 1961. Demographic and Socio-economic Profiles of the Hill Areas of North-East India, Census of India, New Delhi

Roy Burman, B.K. 1995. 'North-East India - An Overview', Keynote Address, in the Symposium on Focus of India's North-East, held at India International Centre on 18th March

10

SOCIAL SCIENCE RESEARCH IN THE N.E. REGION IN THE 21st CENTURY

— *B. B. Kumar*

The 21st century and the new millenium is fast approaching. Many people are talking about the same in a romantic way and are hopeful. They forget that the hope and the happiness shall be brought by the people themselves and their hard work and not by the change of the calendar. The first day of the new century and the new millenium shall in no way be different from that of the last day of the present one. Needless to say that people should work hard and elites should give proper direction for a happy new century. The role of social sciences is crucial in this case.

The North-East region of the country was subjected to reorganization resulting into the formation of seven states. The formation of small states brought the administration and the development machinery nearer to the people. This should have brought development at the grassroots level. Unfortunately, the same did not happen. The development became lope-sided. The development machinery became all-consuming. The percolation of resources at the lowest level gradually decreased. The benefits of development were pocketed by a microscopic minority. The gulf between the poor and the rich widened, The industries and other ventures run by the State Governments and the State Government undertakings, such as, the State Transport Corporations, State Electricity Boards, State run hotels, etc. not only failed to provide revenue returns on the huge investments out of the poor people's money, but continued to run on losses. Non-revenue earning non-productive employment became unbearably burdensome for the common man. However, the emphasise of the State Governments always remained on providing white collar jobs. There was no attempt to overhaul the system of education and the present system of education mostly prepared white collar job seekers. The problem was further aggravated by lack of man power planning by the Government. The researches on these aspects of the development are inadequate and the same is not a very happy state of affair any way. The small states provided immense opportunities of developments which were not adequately availed. The social scientists failed to play effective role in this particular field. The social scientists should have focused more on our development failures and the remedial actions.

The number of books on the states and the tribes of the North-East India published every year is very large. Such publications include the monographs on the communities written in the Loyalist tradition, memories of the retired administrators and politicians, books reporting in the journalistic fashion, edited seminar papers, etc. Many inform; some of the books inform and analyse. There is no dearth of the books which misinform and misinterpret. The micro observations fail to provide macro perception. The balanced blending of the emic and the etic views is often missing in a large number of them. The number of journals, including journals published by the universities of the region, however, is not very large.

The number of publications per year is impressive no doubt; but there are serious gaps. The social science research in many fields has failed to free itself from the colonial hang over. The colonial biases, misperceptions, mis-interpretations and mis-directions continue to dominate. This result into lack of clarity and gives birth to wrong trends. To serve the interest of the empire was the supreme objective of the colonial writers. The emphasis, direction and the trend should have changed in free India. Unfortunately that did not happen. Old trends continued in our ethnographic and historical writings. The freedom and developmental and welfare activities posed new challenges for the social scientists. These challenges were, however, not taken seriously by our social scientists. The states have miserably failed in their development and welfare efforts and we lack serious studies on the same.

The perceptional haziness can not be avoided if we initiate our studies on the wrong presumptions, The myths and the wrong presumptions lead us nowhere. Unfortunately, many of our writings are based on the same. Some times, we pre-suppose something which does not exist at the moment. Such practices need to be discontinued.

The people participate in the democratic process to elect their representatives. Their elected representatives only take the decisions and run the affairs on their behalf. Such representatives may function at various levels, such as at village, district, state or central level. It is the integrity and the capability of the representatives which makes the difference. The united voice of the people forces them to act properly. Otherwise, the representative may become self-seeker, which very often happens today. This human factor is often ignored. The misperception that the political structure and not the human element, shall bring change and proper milieu may be built up by the same, need to be corrected. The expression, such as 'participation of the people in real terms in the decision-making' is, therefore, vague and needs full elaboration.

The Central and many State Governments are trying to attract capital from outside. Their efforts have not been fully successful so far. The states, like Bihar. have failed in their endeavour to attract foreign capital. The West Bengal is hectically trying to attract the same, and to prevant the flow of the capital outside the state. The question is, whether really the capitalist entrepreneurs and multi-national corporations are rushing towards North-East India. If not, then the question of the marginalization of the smaller ethnic groups residing in the peripheral areas and the reckless and unsustainable exploitation of their natural resources does not arise. In this connection, it needs elaboration that our economy failed to grow at the desirabble level due to aristoratic and feudal socialism. Before the Economic Liberalisation of early 90's our growth rate was less then 5 per cent. It is pertinent to mention on that for the most part it was pegged at 3.5 per cent. 40 per cent of our population is below poverty level, and every year we are adding an Australia. How are we going to feed the one sixth of humanity? Foreign Direct Investment (FDI) is, ironically, thrust on this country by the misdeeds of those who are opposing it the most. It needs mention here that the North-East region of the country is going to ask for the foreign investment most, but there is no possibility of its smooth flow due to various well-known reasons.

Our concern for the bright future of the people of the region in the coming century needs the backing of the proper and the planned action. It is not possible to have sustained development, unless the population control measures are seriously taken. The growth in harmony with the natural surroudings and ecological environment is only possible if the population is effectively controlled.

A point, which needs special mention here. and the social scientists must emphasise the same, is the fact that there should be proper harmonisation of the rights and the responsibilities of the people. There should be proper emphasis on the mobilisation of the resources at the local level for the development.

11

IDENTIFYING RESEARCHABLE ISSUES FOR 21st CENTURY NORTH-EAST INDIA

— *Biman Kar*

Concept of Scenario

Like many other dictionaries, The Lexicon Webster Dictionary means by the word scenario, "an outline of the plot of a dramatic work, giving particulars as to the scene, characters, situation..." But the Penguin Modern English Dictionary means by it, "outline of possible future events". According to the Chambers 20th Century Dictionary it means "an outline of future development or of a plan to be followed, which shows this operation of causes and where points of decision or difficulty occur. So It can be taken to mean two things — (1) likely future or (2) a determined future or a wished future. The second reflects the assumptions, value judgement, likings and predilictions of the one who draws the picture. A rational mind may not accept it as a fit subject for discussion. On the other hand, the first is possible when extrapolative exercises are possible and the same is possible when it can be legitimately assumed that certain variables are given and or move in a given direction. Tools of futuristic analyses are available. North-East India is in turmoil, and apparently there is great uncertainty despite that, we can think of identifying researchable issues for rapid development of the region, keeping in mind possibility of only fresh minor geographical delimitation within the broad present frame work.

Militancy and Development

There are two contradictory views (1) It is not so bad, as more concessions, projects, funds etc. can be had from the Centre, (2) it stalls growth and development, as it causes flight of capital, inhibits inflow of private investment from outside the region, disturbs both, the commodity and labour markets, etc.

It is not possible to agree with the first view as appeasement in any form can neither initiate nor sustain development, while a look at economic history is needed for commenting on the second view. Wars, revolutions, conflicts, social tensions etc. characterised Western Europe throughout most of the 19th century. Yet it was during that time that highly industrialised Western Europe emerged. We have recent examples across the globe where militancy, conflicts etc. have vitiated development climate and stalled growth process. The difference lies in whether investment, commodities and labour markets are adversely affected or not. Presently militancy etc. in N.E. India have adversely affected the same markets. In one scenario, militants, insurgents etc. learn from economic history and avoid disturbing the markets and in another they do not. We are not interested in guessing the probability of either. But we find that in the second scenario, an economist can endeavour to spell out what is needed for rapid development, while in case of the 2nd he has to raise his hands. It may be a fit subject matter for sociologists, political scientists and social activists to find out solution to this vexed issue.

Viability Question

At one time it was thought that eventually world's number one position would be contested between China and India, because population size offers vast market and because of economies of scale and effciency. But recent dramatic changes in the art of management and revolutionary changes in industrial and information technology has rendered this notion useless. Both management (long ago divorced from ownership) and production can now be profitably widely dispersed. It is no longer compulsory to maintain a vast network of branches either in buying or selling. Smallness of size of a country need not necessarily be an inhibiting factor. It may be thought that if imperialism explained affluence of even tiny countries in the past, the above would achieve the same in future. At least that is the faith of the globalisation ideologues. But the emerging scenario in market capitalism (the process of increasing concentration of economic power and hence decision making on earth would further strengthen increasing flexibilisation of labour markets, would marginalise social security network making working class more and more vulnerable, smaller the developing State, weaker would be its bargaining strength in international competitiveness etc.)— making it necessary to find out the viable size of administrative units. Research is needed to find out solution to ethnic aspirations to some alternative other than miniaturising State or administrative units within a State.

The Question of Population Size and Composition

A major field of research in N.E. continues to be immigration dominated high population growth rates and its various features. The problem of reducing or threatening to reduce the indigeneous people into a minority in their "natural home land" by the immigrants from outside, including those from across international border is a serious one. There is the emotive issue of giving citizenship right to the refugees from Chittogong Hill Tract (1965 or so) in Arunachal Pradesh.

While indigeneous people cannot be allowed to lose their identity or their traditional command over productive resources, undue restriction of productive inputs and technology can also not be, encouraged. On the other hand corrupt practices like illegally leasing out land or industrial license have not served the purpose of statutory restrictions on outsiders well. Research is therefore needed to find out appropriate means to protect tribal and indigeneous interest without sacrificing socioeconomic development. Effective but not prohibitory restrictions need to be placed on incoming productive inputs, so that, tribal interest is not sacrificed.

Population pressure on productive resources and ecology is another area of great concern for our region. Official and tradition malpractices coupled with population pressure has severely reduced the natural forest cover in N.E. Assam and Tripura are probably worst affected in this respect. Research is urgently needed to identify how to retain needed forest cover and to resettle those who now illegally converted forest land into human habitation.

North-East Specificity in Research

Our region suffers from a number of handicaps, much of which have already been researched into. There are considerable data gaps. This is more so because of immense variation within the region. On issues that are common with other parts of the country, lot of good research has already taken palce elsewhere. Such issues as non-implementation of labour laws or land reform measures or child labour laws or organizational weakness etc. should not top priority list of researchable issues for 21st century. N.E. specific issues should get top priority. Ethnic segmentation of labour market (semi skilled and unskilled category) is one instance. On the other hand ethnic question often seem to overshadow economic consideration. A healthy and strong labour and trade union movement for achieving just cause for working class need an integrated and not segmented labour markets. Again strong trade union movement demand immersion of such differentiating identities as races, language or religion. Research is needed to find out as to what extent such factors have influenced trade union movement and how to remove them. To take another example, even though notionally women in matrilineal societies enjoy higher status, they in reality have little role in decision making process and may be, have to work harder than their non-tribal counterpart. Emancipation and empowerment of women is an essential component of development strategy. Such problems which are specific to out region and hence should get priority in our list of researchable issues.

Institution

Several concepts, each deserving merit, when studied in isolation, do not seem to be conducive to growth and development, when studied together. Thus there are concepts of "civil societies", traditional self-governing institutions, Panchyati Raj Institutions, Autonomous District councils, traditional or customary laws etc. Research is needed to harmonize and reconcile them, so as to make them agents of development. Kerala offers a good example of mass mobilisation and participation in rural development process through P.R.Is. Research is needed to find out the ideal grass root level institution specific for each region or sub region, which, while protecting and preserving valid interests of the indigeneous population, can also become an engine for development.

On the other hand, while invasion by market extension and administrative expansion or population influx threatens the identity and economic interest of small communities, inter-tribal differences or feuds prove inimical to growth. Research endeavour is needed to find out some such solution as formation of tribal confederation, which while settling inter - tribal disputes and preserving tribal identity will ensure removal of all social and cultural impediments to development.

Liberalisation etc.

However firm our reasoned conviction in socialism may be we have to accept liberalisation and globalisation as unstoppable. If the glorious days of "Public sector" proved short lived, the present wave in favour of the new paradigm will also be transitory, because increasing disparity both between and within countries, increasing fragility of working class and hence the weakening of purchasing power of the masses, is bound to make free market mechanism, unsustainable in the long run. Meanwhile poorer and weaker regions within developing countries may have to face backwash "effects" of the new developments, than other regions. So for our region, research is needed to find out strategies to minimise the backwash effects. In our humble opinion, taking advantage of traditional communal ownership, co-operativisation would be a good means to overcome organisational and bargaining weaknesses. Relocation of very sparsely inhabited areas is another need. Research is needed to find out how ends can be achieved cost effectively.

12

THE NORTH-EAST AND FUTURE TRENDS

— *K. S. Singh*

As the Twentieth Century draws to a close the North-East emerges as an intensely researched region in social sciences including social history, with a strong local scholarship making its mark. This rise of local scholarship and the insight provided by is intrinsic to the understanding of the people and the framing of appropriate strategies for integration and development. Since Independence the North-East has contributed immensely to the understanding of plurarism, autonomy, ethnicity, "Specificities" or in other words of people's aspirations, devolutions etc. The people of India project for the first time surveyed all the communities of the North-East generated an enormous amount of data on culture, languages, and biological variations of populations This new data set supports the widely held view that the North-East is emerging a composite region ethnographically where communities have began to interact and share a great deal; Politically as various forms of experiments are being conducted in autonomy, and as importantly as a unit of planning and development. Gender issues in the North-East are being highlighted for the first time. Lastly, the importance of the North-East as commercial and cultural bridge to South-Asia and South-East Asia is being recognised as never before, and appropriate policy measures are being initiated to use this to advantage.

In the next millenium these trends will continue and some of them will even intensify. The various identities, ethnic, regional etc. will consolidate themselves as search for roots, for scientific, history, for healthy relationship with one another will continue. As the ethnic situation stabilizes and inter-ethnic conflicts will subside, the communities are barred to turn to the intensive exploration of the past and their "rootendness" in pre-history, archaeology and anthropology. The North-East will be finally recognised, as DNA based studing of migration of populations in pre-historic period advar.ce, as the transit routed taken by the populations migrating to South-East Asia.

Another subject being focussed is the rich rapertoiric of folklores, which will be compiled and interpreted to understand the people's inner world, their psyche closely linked with their identity. Last but not least, is the ecology and environment of the North-East, which is recognised as one of the richest areas of biodiversity. The complex relationships of the communities and their environment now under stress need to be explored both intensively and extensively and placed on a sound, healthy basis. Traditional and appropriate technologies being identified and even applied should help the North-East exploit its enormous resources and increase its productivity in order to create an economically advanced region and a culturally stable system. The links with the rest of the country should grow as mutual understanding and mutual respect develop further with greater interaction in all fields.

13

ADULT EDUCATION AND RURAL DEVELOPMENT

— *Meghali Baruah*

Development is a process related to the development of human beings in every walk of life. Primarily, it is concerned with the development of individual and then form individual it goes to the community, mean that every one is given an opportunity to realise the potential to stand on one's own feet and meet one's basic demands. Development is not just growth, but growth centred around man for him to live in harmony with his environment with a sense of values, of human dignity, and with self-reliant and self-confident attitudes. Development is an integrated whole. It does not start with goods, it starts with people and their education, organizations and discipline. Without these all the resources remain latent and untapped.

Education has long been recognised as the nucleus of an overall national development. Any comprehensive approach to development underlines the significance of education in three related ways - as a basic human need, as a means of meeting other basic needs and as an activity that sustains and accelerate overall development naturally. "One must think of education, not only as a sector development parallel for example to agriculture or industry, but also a pervasive element that must be integrated horizontally and vertically into all development efforts". Education is not only one of the key indicators of socio-economic and political development, it is also a major component of human resource development.

In India, literacy has been a national priority ever since it became apparent that illiteracy is one of the major bottleneeks in India's development. With little access to information, modernisation and communication and virtually no access to minimal survival tools the non-literates remain helpless victims of poverty, deprivation and exploitation. There has been a growing realisation among policy makers and planners that development will never become self-sustaining unless there is a change in the attitudes, values, knowledge and skills of the people as a whole. The only way this can be accomplished is through education. Literacy is virtually the pre-condition for an individual's evolution and national development. A literate society can ushar in progress and prosperity quickly.

India has placed poverty alleviation and rural development higher on the national agenda. This has led to the launching of various programmes like IRDP, JRY, TRYSEM and DWCRA.

However, not all these programmes have been an unqualified success because the beneficiaries who are the poorest of the poor are unable to appreciate the schemes and their objectives. Being unlettered, they are often cheated out of their rightful share of subsidy and loans.

It has now been universally accepted that developmental efforts like family welfare, sanitation, health and agriculture are bound to fail if the population is not literate. Therefore any developmental efforts should be linked with education of the rural adults.

However, in a country like India which has the world's second largest population (860 million according to 1991 census) and the largest number of illiterates (330 million according to 1991 census) it is not a mean task.

While literacy percentage in India has grown at a rate of approximately 8 per cent per decade there has been no substantial decrease in terms of the number of illiterates. India had over 300 million illiterates as per the 1951 census and even after the literacy rates have risen to 52.1 per cent in 1991, there are 300 million illiterates. Of these, over 110 million fall in the age group of 15-35 which is the productive and reproductive age group and the target of India's developmental efforts. India has been unable to effectively contain its population and the population growth of 1.8 per cent is still. too high.

The population problem has other dimensions. There is a wide gender disparity between male and female literacy rate. While the male literacy rate stands at 64 per cent, the female literacy rate is a mere 39 per cent . Similarly, there is an imbalance between the rural Urban literacy. While 70 per cent of India's population resides in villages the rural literacy rate is a mere 44.69 per cent as compared to the Urban literacy rate of 73.08 per cent. The female literacy rate in rural areas is even lower being only 30.62 per cent.

India's women, the Scheduled castes, scheduled tribes and other backward classes, which are the under privileged sections of the society, are the target of India's adult education programme. As a majority of them live in villages, the literacy efforts have an impact, direct or indirect on rural development.

Adult education was not high on the national agenda immediately after independence. It was only from the first five-year plan onwards that adult education was linked to rural development. Consequently, the NAEP was launched on 2nd Oct, 1978 which has now been included under the broader umbrella of the NLM which was set up in 1988.

It is one of the most concerted effort so far to promote literacy in the country. It is based on the premise that literacy is an indispensable component of human resource development. It aimed to harme the advances in science and technology and in communication and pedagogical sciences for the benefit of the society. The NLM today imparts literacy to persons in the 15 to 35 age group through compaign mode. It has established important milestones by bringing 448 districts under the fold of literacy campaigns. The coverage in terms of targeted number of non-literates exceeds 130 million. Approximately 72 million people have been made literate.

The NLM seeks to impart "Functional literacy" to non-literate. Functional literacy implies imparting of skills of the 3R's which have a bearing on a person's life. The environment building stage which is crucial to the success of the literacy campaign emphasises this aspect to the illiterates who invariably asks the question "Literacy for what"? The truth is that in India literacy is not the felt need of the people. Employment is . Education is still considered to be a luxury for the poorer section of the people especially in rural areas. And only if literacy can be seen to lead to the economic betterment of the people, it has a chance of success.

Adult education, therefore emphasises on functional literacy which will improve on individuals productivity as a worker and enable him to take up a trade or a vocation. Keeping in mind this factor, the priments have emphasised saveral aspects of rural development like better agricultural practices, consolidation of land holdings, rural credit, equal pay for equal work, women's rights etc. It is belived that proper transfer of these knowledge would definetly benefit the rural Poor.

The total Literacy Campaign which are implemented through Zila Shaksharta Samiti exemplify an interactive and a communicative process of management and implementation. The objective of the TLC is to improve the overall quality of life. In order to achieve this objective the unique strategy undertaken is environment building which is basically carried out in the form of Kalajathas, Padyatras, Slogan writing, street plays etc.

But it is at the post literacy stage that links between literacy and rural development are sought. This link can be carried through 1. Dissemination of information-about the various rural development programmes through post- literacy material and awareness creation and 2. Organisation of the neo-literates with the aim of carrying out-economic activities in such a manner so as to result in the conferment of ownership rights of assests to them so that they become member of the collective.

The success of adult education depends entirely on the post literacy programme which aims to maintain continuity of education of the adult learners. It is at this stage that literacy is linked to rural development. The purpose of this stage is to produce a series of literature to enable the rural neo-literates to take benefit of the latest developmental practices.

The experience of the post literacy programme in Assam like in other parts of India is not uniform. It varies from district to district depending on the imagination, commitment and capability of the people involved. Thus, while there are pockets of success where communities stand invigorated after the literacy campaigns, there are also districts where the campaign flounders after the basic literacy phase because, the post literacy campaign does not capture the imagination of the neo literates.

A dearth of good functional literature in regional languages is also a constraint. While the state resource Centres do bring out books on various aspects of rural development most of the literature is available in Hindi or English and not in the regional languages. Moreover, there is a misconception that any book can be given to anyone who can read and that proper reading habits can be cultivated in due course, But this is not the case with neo-literates. Many books printed and published for the neo-literates are quite in appropriate and even curb the reading interests of such people . A neo-literate book should be such that at first sight one should be tempted to take the book and examine it, read it at a stretch and read it again and again for ensuring better comprehension. The content should be such that they got some thing new from their exercise of reading a book.

Another problem is that, while vocationalisation is a major thrust of literacy campaigns, there has been no systematic effort to link It with learning needs of the neo-literates as has been done in some other countries, while women in the post-literacy phase are taught stitching, there are usually no manuals on stitching and no books on dress-designing for non-literates which would not only enhance their skills and vocational skills are compartmentalised without there being a constitious effort to strengthen the vocational skills through the written word.

Moreover, there has been no link between the post literacy campaign and the various agencies of rural development. The neo-literates feel that they are unable to benefit from their learning experiences in procuring help from these agencies since they find the procedures very complicated. Therefore, to make this movement a success there should be a concerted effort from these agencies also. Various developmental egencies should make it a point to constantly feed the adult neo-literates with latest development in various sectors, but in a very simple manner. Actually the fact is that, the post literacy programme depends entirely on the harmonious cooperation of the various agencies meant for rural development. For this purpose, these agencies are to be sensitized to this issue.

In this regard, the role of the universities of the state can not be ignored. Various programmes of adult education has been undertaken by the Dibrugarh and Guahati University as part of their extension programme, but they are not upto the expected level of success. Serious efforts are to be taken up by the Universities. It should always be kept in mind that teaching, research and extension programmes have been accepted as three supreme goals of Universities. Although teaching and research have become full priority of the universities, they have somewhat forgotten the extension services. A broad programme of adult education can be carried out by the universities and this have to be taken up with much seriousness.

In conclusion, it can be said that in order to ensure that the gains made in course of the literacy and post-litteracy campaign are not last, the NLM will necessarily have to plug the existing lacunae and loopholes and help in the creation of a learning society whose members will enjoy learning for its own sake, for their economic betterment, for community participation and collective action which in turm will foster a dynamic model of rural development.

14

ENHANCING TEACHING AND LEARNING INTEGRATED SYSTEMS ANALYSIS

— A.S. Guha

New Strategies in Education: Openness and Flexibility

Today, educators as an extension to classroom teaching must also posit the role of facilitators in teaching and learning. A teacher is also to facilitate learning, organize not only his/her teaching mannerisms but should also be a helper in individuated learning processes. Extension of the classroom is a must whether it is organizing tutorials, holding inter-cultural programmes or even informally meeting students. In fact, teaching research and extension should be the focus of all academic activities. To facilitate learning arguably calls for innovations and experimentation. Developing good quality study texts should be a priority. Unfortunately very few academicians give importance to writing standard text books. A research paper produced by a teacher is not only a document to be mentioned in the bio-data but something which ought to be shared among students giving the wider scope to be discussed critically in the classroom. Abstruse and 'difficult' subject matters should be simplified so that students can readily grasp the underpinnings of complex theoretical abstractions, which in turn may help for future research.

Extension in teaching and learning could mean then the use of audio-visual contraptions and overhead projectors. The underlying basis of teaching is that of a sensory experience both for the teacher and the taught whereby communication processes are two-way and not simply one-way. The interactive methods of teaching can further be supplemented by using diverse formats such as group exercise, quiz, debates, seminar presentations etc. where also switching from one format to another Is experientially rewarding academically ; both for teacher and student alike. In this manner too the classroom can be 'extended' to other components and formats in which peer group activity is another aspect of classroom activity, or is a miniature classroom. The much talked about student participation can be actualized by such extension based work in the teaching/learning exercise where role play is a categorical device.

Facilitating learning in a broad and innovative sense takes the 'teaching' out of the classroom, so that the 'chalk-and-talk' theory is neatly subverted into some more fundamental processes of teaching. A student must be able to create his/her own resources of learning in addition to lecture notes and text books. This could be the use of T.V., radio and audio visuals for learning; planning of study hours can be judiciously made, relating this actively to course work.

Evaluation in examinations has been simplistic where the marks or grades awarded are beyond question. Proper justification of grades awarded, strengths and weakness of an answer supplementary references etc., are a must which also require skilful academic handling and expertise. They are in their own right an integral part of teaching methods.

A teacher's role does not begin and end the 'classroom'. He or she must possess the vision to organize as many 'classrooms' as possible so that teaching and learning is dialectic and are discourse based activities. A teacher is an organizer, an innovator and an academic administrator in the final reckoning.

To think of the 'disadvantaged' in educational contexts is to invest also a broad continuity to the processes of learning where educational planners and decision makers are also providers for a life long learning experience. The disadvantaged are those who want to learn and always do not get the opportunity to do so. In such situations educators must be continually proactive in terms of planning and strategy. The term disadvantaged does not only have economic implications. The word is all embracing, encompassing all aspects of social, economic and cultural realities. But a continued effort to search let us say for education in one's life which is translated into a life long experience will benefit social reprobates, 'outcastes' and of course professionals who will be able to 'sandwich' their studies between the learning time and the working time. Work and the job realities are so much of a prepossessing factor that even if there is the intrinsic desire to learn, extenuating circumstances of profession and work compel a person to pull out of studies. In a highly competitive ethos a common language of study is essential but bilingualism as a means of study, will be of greater help to multilinguistic and multicultural societies. However, we must understand, in this context the needs also of globalization, globalization and cultural identity are complementary in a sense, they should not be necessarily viewed as a clash of ideologies, and even if there is a conflict of such values there is a need to reconcile them. The disadvantaged may not have the privilege of being educated in the language of the elites or the 'common' language, so multilinglialism while offering courses or the indigenization of course development is a must in binary cultural situations. Course material translation from one language into different languages will drive home such essentially in pluralistic- societies. Today as we enter the twenty first century there is a rising internationalism but at the same time there is search for specific roots. Such tension generates contradictions in societies and those who are disadvantaged for casual factors such as social, economic, inequality between sexes etc are caught in such a vortex of change and contradictory pulls. They must be given the chance (learner autonomy) to study courses of their own choice, in their own language and to register for such courses at any time of the year. New 'need' based courses specially designed for say the physically and mentally challenged, drug addicts, working women, people living in geographically remote areas where economic sustenance is severely limited should be identified on priority basis. The emphasis is also on training as part of life-long education. Inter cultural courses In multi-ethnic situations as for example India's can help people in remote locales to be aware 'of heritage common to all humanity".

It is also the question of shared values and the inherent desire of a life-force to 'live together' which can wean away the disadvantaged from the quagmire of exploitation and the brutalization of society. Shared values will also mean :

— Opportunities to learn
— To learn in a language of one's choice
— Freedom of choice (of courses)
— Opportunities to be trained during work period

— Opportunity to work and learn
— Opportunity to be 'literate'

Underemployment and unemployment causes the breakdown not only of family life but of society as a whole. "A hundred and twenty million people in the world are unemployed and more than one billion live in abject poverty". The marginalized and the isolated need empathetic attention so that education and distance education in particular can integrate these with the 'mainstream' of society.

Traditional knowledge and education has focused on abstractions rather than on imagination, communication and leadership skills. The challenge is : how do we promote academic excellence in such a manner so as to discover these immeasurable realities, a sense of beauty, sensitiveness to one's environment or say even the development of manual skills ? Can distance education promote the aesthetics of education or intercultural education ?

Education in the 21st century must combat all forms of exclusivism. An exclusivist educational pattern such as rejecting students because of one-time academic failure may lead to social aberrations, and the emergence of 'drop-outs'. School and College dropouts are disadvantaged people but distance education methods will have to cater to such disadvantaged youth in-terms of short-term and vocational modules. The rural youth for example need special attention and consideration in this regard. Devising forms of education for different minority groups both nationally and internationally needs to be given priority.

Distance education provides wide scope for renewing and refreshing our approaches to educational systems calling for absorption and eclecticism. It gives wide currency to innovation, change and flexibility as there is an extension of knowledge based realities into their proper relevant and adequate applications. It is also extension based studies and application oriented knowledge. There must be a fine distinction between information societies and learning societies. Adapting one's self to a rapidly changing world is not only the prerogative of the educationally 'qualified' but also that of the educationally 'backward'. The inequalities too between the sexes in the sensitive field of education brings discrimination. The girl child for example in countries such as India is already destined for marriage as an axiomatic and apriori truth. Priorities are wrong, what about education for the girl child?

Basic education for everyone is one of the fundamentals if not the fundamental of teaching and learning. The jobless youth left to fend for themselves in inimical surroundings are a potential source for endangering national security.

A monolithic model of education can be developed by reducing the distinction between initial education and lifelong education. This hiatus can be bridged by Distance Education applications. The different layers of education must be perceived holistically so as to integrate them into one complex whole forgetting terminologies and description of educational heuristics. Distance Education is also a facilitator of teaching and learning not loth to retaining old practices if found suitable; in a given framework. The local circumstances of education, the indigenous needs of it are of crucial significance. In a manner of speaking all sections of the society may be 'disadvantaged' in one way or the other, or at one point of time or the other.

As we have entered the 21st century, what is discernibly evident is that the philosophy of education and its applications area undergoing sea-changes. In fact the very concepts of education are radicalized and revolutionized to introduce newer contexts, settings into what may be described as the 'extended' classroom.

This is largely due to the fact that technological processes in education have exacerbated innovative thinking in this field. The applications of technologies here have been manifold beginning from radio/T.V. broadcasts and telecasts, use of audio-video cassettes to audio video conferencing and on-line computer teaching. The 'classroom' thus has been recreated, extended and ramified to cater to the work force, service clientele and different professional groups. This again is mainly and preponderantly due to the ontologies of Distance Education methods of teaching. However, the word of caution and introspection here is that technology also has its own appropriateness depending upon the social and economic development of a place or region. We can then think of disseminating knowledge to underprivileged and disadvantaged groups of society.

A 'holistic' view of education caters to vastly heterogeneous groups as well as traditional and non-traditional categories of learners. Thus experimentation is needed in even conventional and traditional set ups. The idea would be to gear towards non-formal and vocational courses, and towards more openness and flexibility even within an existing framework. This will lead to barriers being broken keeping in mind professional growth and definite target groups.

Openness and Flexibility, The Case of St. Anthony's College: St Anthony's College Shillong has achieved exactly the goals and objectives sought for above. The college has innovated within a formal set-up by introducing non-formal and vocational courses. In this manner it has broken tradition while at the same time retaining it to the greatest extent needed. The college has undoubtedly introduced open and flexible norms within an existing and patterned set-up. Such innovativeness is certainly unprecedented in the northeastern region and should be a trend settler. Granting autonomy to the college will create more flexibility and innovative thinking.

Total Learning Situation : Learner Autonomy and Learner Contiguity

Learner autonomy points to the learner centric vision of education, its learner centredness and the appropriate methods adopted to achieve it. More and more the concept of education has reached a point, where we enumerate more positively learner roles rather than teaching roles. This is the pedagogical shift that we envisage and posit emphatically in teaching-learning methods. The cultural shift that is taking place is that teaching is mediation; parallel texts and sub texts are entering the teacher - learner classification. These are parallel to the methods already adopted in classroom teaching, i.e. the popular but not sacrosanct notion of the lecture method.

The cultural shift ᐨeferred to the above is part of the paradigmatic shift that is evolving into let us say new teaching-learning systems. It is the purpose of this discussion to examine such 'newness' to invest a philosophical basis to the concept of education. What are, who are, the mediators in the teaching/learning process ? What is the relevance of target groups here ? Can education be subversive, i.e. can it break the barriers of school education, higher education and university education to apprehend contiguity and wholesome features.

The mediated form of instruction by way of study texts and technology aided instruction makes the educational experience lively, interactive and independent. Home - study for example, the opportunity to study at one's place and pace is part of the objectives of continuing and adult education, where a situation is opportunistically created in which those who have missed out are given a kind of flexibility to learn to one's advantage and benefit. The learner is then not an anachronistic idiot but the creator of an ambience, a cultural situation, where the learner is : A Teacher' and 'A Trainee.

Let us examine the points mentioned above nonchalantly one might think, which may be (mis) represented as a lethargic and lackadaisical view of education. For too long the concept of education in India has remained in the morass of rigid stratification : the teacher 'teaches' 'the learner 'learns'. The teacher is omniscient, an intellectual glut and a powerful' (but of course superior!) disseminator of information. The student - learner relationship is thus construed chiefly and pedagogically as linear and not causal. That, there may be a cause-effect relationship between the teacher and the learner is glibly overlooked. This is the cause—effect relationship to which I am looking towards a pointer; towards 'new' teaching-learning modes, the roles are reversed, the teacher is a learner, the learner is a teacher. The optimum use of technology aided instruction which is interactive, constantly shifting gears let us say, can achieve this.

Learning is then virtual and autodidactic, classrooms are then created at homes, in trains, in buses anywhere. I am thinking of culture specifics: of the Indian situation, where the rigmarole of the educational scene has become nightmarish - learning by rote and cramming, three hours 'essay' (whatever that might mean), questions, etc. The learner is a trainee because the learner has the flexibility and the right to : 'Choose to learn' and learn to choose'.

The learner may be in a position to select specific, target-oriented courses, and relates them to professionally objectivised goals ("Learning to choose"). If we are to design courses for laboratory technicians we are designing a 'professional' programme just as that of an MBA programme, or a Masters In Computer Applications programme which are very popular in this searingly competitive world, in what we try to capture as a professionally competitive ethos, such a tag also becomes a misnomer in third world countries. Learner autonomy and academic freedom should express more independence and the freedom to choose, a housewife should be in a position to pursue a programme in creative writing and a journalist in environmental studies, but who will good them, will it not be a question of autonomy and autodidacticism, that of learning for 'disinterested' purposes? The creative will is energising but also self fulfilling, jobs and professions come and go but the interest to learn and the acumen for it is perennial.

We have addressed learning points of view. What then of teaching, how can 'classroom' teaching be extended into creating and multiplying a larger number of them ? Let us begin with evaluation, evaluation should be continuous, both formative and summative, evaluation should be innovative, there should be lab intensive work, extended contact etc. Tutorials should exercise maximum options such as group discussions, brainstorming sessions, quizzes, small group interactive sessions etc. The teacher then is also a coordinator and a manager. This mediating role of teaching can then perform wide ranging functions from tutoring skills to administering and managing.

The open university experience however has wide ramifications on the Indian educational scenario especially after the establishment of the Indira Gandhi National Open University by an Act of Parliament in 1985. It radicalized teaching-learning methods and the evaluation systems, it used the concept of technology with expediency and alacrity matching the print media with audio video components and updating such heuristic skills with audio/teleconferencing. It is now in the way of marginalising conventional/classroom/face to face teaching integrating featuristically and futuristically the print and the electronic media; tells is surely an attempt at upgradation of classroom systems, methods and practices. The university has proliferated into more adaptable practices of facilitating learning evolving practicably, concepts of learning facilitators, specific programme centres and work centres. This also brings us to the question of academics visualising roles of facilitators to envision a total learning situation. The functioning of an education system in the complex form of an industry projected the 'business' environs not anarchic and profiteering, but cost effective and cost reducible. This has led to the university (IGNOU) in collaborating with partner business institutions and multimedia teaching Institutions where the state of art of technology is used for teaching courses in the active and dynamic field of management and computer sciences.

However to make this education system more mass based, invigorating and fructuous professionally related academic programmes have also been introduced in the spheres of teacher training, higher education, food and nutrition, rural development, journalism and mass communication health education, creative writing etc. This spirals into an educational ethos both vertically and horizontally where a varied and complex target group with different professional backgrounds and compatibility are taken into full cognisance.

In India now the vision of higher education must be constructively understood not as compartmentalization (school education, college education, university education) but as a holistic means to achieve objectives of TLS (Total Learning Situation). To illustrate my contention the courses cited above in terms of diplomas and certificates may help to rationalize such goals. Extension studies in areas such as agriculture, horticulture, disaster management, art and craft, health and sanitation are other areas where education is mass-based, an awareness; of needs, of surroundings. Correspondence course Institutes, Distance Education Directorates and Open Universities must be integrated meaningfully, resourcefully and pragmatically.

The ends to teaching and learning must also be creatively envisaged as expansive, changeable and adaptable; shaped accordingly to the demands of hungry learners whose thirst for knowledge is unassuaged. Education and the attendant progress made with it is pluralistic and complementary, evolving from total literacy, total literate(acy) to a total learning/teaching situation, Indeed a sensitively tuned ambience, in fine accordance with the spirit of continuing, nay; Continuous Education.

Conclusion : Integrating Diverse Areas of Knowledge with the New Paradigm

The 'new vision for learning' envisages perforce the concept of learning to learn (together). The UNESCO World Conference On Higher Education (1998) posits that institutions of higher learning must play a supportive role in identifying and addressing issues that affect the well being of communities.

What are learning organizations ? The education system must constantly aim at a synthesis of 'new' methods with 'new' issues that affect community well being (Joshi & Puspandham K. 2000). For example the National Council For Women's Education in 1964 recommended that sex education be imparted at the middle and secondary stages of school or collegiate education. The National Committees and Commissions have focused on a multidimensional approach to bring about positive directional changes in Women's

Education in India. Some of the thrust areas are : developing a positive attitude towards women in curriculum, encouraging participation in vocational, technical and professional areas, promoting research centers of women's education in different Universities and setting up organizations like National Council For Women's Education. In short there must be a positive Interventionist role to be played by the government (Singh, Nisha 2000).

It is in this context that the question of extending the classroom arises. "The destiny of India is shaped in her classrooms" (Education Commission 1996) Integrated new technologies with teaching methods posits a holistic vision of education beginning with literacy at the grassroots level to higher education ; the emphasis being on continuing and lifelong education. To establish a synergy between technology and education is to autonomize it. (Ghanci, D.A. 2000). The industry university alliance is another case in point. UNESCO has designed a programme known as University Industry Science Partnership (UNISPAR) with the following objectives :

- — To match university education with industry needs
- — To transfer research findings from university to industry
- — To continuously, reorient university curriculum to cater to every developmental sector
- — Prompting academics and industry to move to each other

There must be a synergy between good academic institutes and good industries. (Naik, B.M. 2000).

In the overall framework of higher education in India there is an urgent need to design courses for those who live in semirural, rural and backward areas. It is pertinent to observe that about 30 per cent of the higher education institutions in the country are situated in these areas What can Distance Education do to play interventionist roles in this regard is the crux of the matter ? (Vernal, Louis 2000).

Distance Education can lead to the spread of population education, it can also be used as a powerful instrument to eradicate literacy and promote informal education. This is the holistic vision of education which I anticipate ; total and composite, where teaching and learning is enhanced through an integrated systems analysis (Kalavathi, P. 1991).

REFERENCES

Delors Commission Report; UNESCO, Paris (1997)

Ghanci, D.A., *University News*; Vol 38 No. 17 New Delhi (2000)

Joshi & Puspandham *University News*; Vol. 38 No. 12, New Delhi (2000)

Kalavathi, P., "Foundational Approach to Teaching Learning Process In D.E", Doctoral Work; Osmania University, Hyderabad (1991).

Naik, B.M., *University News*; Vol 38 No. 17 New Delhi (2000)

Pandey, Saroj, *University News*; Vol 38 No. 21 New Delhi (2000)

Singh, Nisha, *University News*; Vol 38 No. 7, New Delhi (2000)

Vernal, Louis, *University News*; Vol 38 No. 19 New Delhi (2000)

15

HEALTH AND NUTRITIONAL STATUS OF THE PEOPLE OF NORTH-EASTERN REGION

— *Pranay Jyoti Goswami*

Introduction

Health is fundamental to human progress. People in sound health can accelerate the pace of economic, industrial and social development.

The Directive Principles of the State Policy of Indian Constitution mentions. "The state shall regard the raising of the level of nutritions an standard of living of its people and improvement of public health as among its primary duties....". The constitution of WHO says, "enjoyment of the highest standard of health is one of the fundamental rights of every human being without distinction of race, religion, political belief, economic and social condition.....".

The Alma Ata Declaration (1978) of which India is a signatory, has emphasised the "People's right and duty to participate individually and collectively in the planning and implementation of health care programmes. "National Health Policy (1983) also emphasised "Health for all by 2000".

After independence, as a result of various immunization programmes like cholera, small pox, poho and some of the programmes like Malaria Control, Leprosy eradication, tuberculosis control etc. under-taken by the Government of India, the death rate in the country has declined to 8.9 in 1997 from 27.4 in 1941.51.[1] The infant mortality rate considered to be an index of health status, has shown a steep fall from 129 per thousand live births in 1971 to 71 in 1997. The life expectancy at birth has increased from 45.6 years in 1971 to 62.4 years in 1996.[2]

Inspite of these achievements, a large number of people are living below poverty line; they are not getting minimum calorie also. Due to various factors infant mortality rate, child mortality rate are alarmingly high.

It is now confirmed that 'Health for All by 2000' is not feasible within this time limit. Govt. is now thinking to reevaluate the National Health Policy and set the time limit to 2010 for Implementation of the programme "Health for All".[3]

Health and Nutritional status of the people of North-east.

Table 15.1 reveals that in 1983 the Birth in the N.E. States ranges in between the lowest of 23.1 in case of Nagaland & the highest of 35.4 in case of Arunachal Pradesh against National average of 33.7. The death rate ranges in between 6.3 in case of Nagaland to 17.3 in case of Arunachal Pradesh against National average of 11.9. In 1992 Birth rate ranges in between 19.2 in case of Nagaland to 30.8 in case of Assam against National average of 29.2. Death rate was also below the national average 10.1 in all the states except Assam (10.4). In Nagaland both birth rate and death rate was the lowest. In 1992, the death rate in Nagaland was only 3.8. Infant mortality rate was also lower than All India average (72) except Assam in 1998. Total fertility rate was lower in all state except Tripura (3.9) and Meghalaya (4.8) than national average (3-3 Table 15.2).

Table 15.1 Estimated Annual Birth Rates and Death Rates (Per Mille)

State	*1983*		*1992*	
	Birth rate	*Death rate*	*Birth rate*	*Death rate*
Arunachal Pradesh	35.4	17.3	26.6	9.4
Assam	34.7	12.1	30.8	10.4
Manipur	28.0	12.1	19.5	5.6
Meghalaya	30.0	8.3	29.8	8.5
Nagaland	23.1	6.3	19.2	3.8
Tripura	23.7	6.3	23.1	7.8
All India	33.7	11.9	29.2	10.1

Source: Basic statics of NE Region, 1995

Inspite of these achievements area-wise variation is also there within the North Eastern states. A large number of people of North eastern states are living below poverty line (see Table 15.2). In

Table 15.2 : Demographic Indicators of the States, North-East Region, All India & Kerala

State	*Population Size in Million As On 01-03-99)*	*% of Total Population*	*Total Fertility Rate 1997*	*Infant Mortality Rate 1998*	*Contraceptive Prevalence Rate 1999*	*Population Below Poverty Line*		*Female Literacy Rate*	
						1983	1993-94	1981	1991
Manipur	2.21	2	2.4@	2.5	20.1	12.3	33.8	34.67	47.60
Arunachal Pradesh	1.2	1	2.8@	4.7	14.0	27.1	27.1	14.02	29.69
Assam	25.6	2.6	3.2	78	16.7	23.5	40.9	NA	43.70
Tripura	3.6	3	3.9@	49	25.2	23.0	39.0	38.01	49.65
Meghalaya	2.4	2	4.8@	52	4.6	28.0	NA	37.17	44.85
Nagaland	1.6	2	1.5@	NA	7.8	27.1	37.9	40.39	54.75
Mizoram	-	9	.09	NA	23	27.1	NA	68.61	78.60
India	981	-	3.3	72	44	37.4	36.0	29.85	39.19
Kerala	32.0	3.3	1.8	16	40.5	NA	NA	75.65	86.17

@Three year moving average TPR 1995-97

Source: Statistical outline of India 1988-89. Tata Services Ltd. Department of Economics and Statistics, Mumbai.

Economic and Political Weakly, May 21, 1994, p. 1305

Yojana. August 2000, p.68

Journal of NEICSSR, Oct. 1994, Shillong p. No.40

some of the states like Assam. Tripura, Nagaland, the percentage of people living below poverty line was above the national average (36) in 1993-94. In the hill areas of the North-east, more particularly in the rainy season, the people have to face food shortage. Due to disruptions of road transport and communication facilities. the starvation and miseries of the people go beyond imagination.

It is known to us that lack of access to right foods, irons, proteins and micro nutrients such as iodine and vitamins is the principal cause of very high incidence of nutritional deficiency diseases; anemia, diarrhea, night blindness, goiter etc. Low body resistance due to malnutrition can complicate minor ailment and make it a health hazard. It can often obstruct or nullify the supportive medical treatment of a child, say in case of diarrhoea.

From the NSS report 238/1-11 (source-EPW, August 25, 1984) it reveals that except Assam average consumption of protein in rural areas of all the North-Eastern states was higher than national average of 76 gm. However, the rural people of Manipur and Meghalaya were getting less calorie than national average of 2724 K.Cal; but per capita consumption of protein per diem in the Urban areas of the states of North-east except Assam & Meghalaya was higher than national average of 75 gm. In the Urban areas of all the states of North-east India calorie content of food items consumed per diem per consumer unit was higher than the national average of 2539 K.Cal (Table 15.3). From this we can come to the conclusion that the nutritional level of the people of North-east India was not below the national average.

But NSS 50th round data reveals that per capita per diem intake of protein in the rural areas of all North eastern states was below the national average of 64.2 gm. (Table 15.4).

If we compare these two reports, it reveals that the per consumer unit per diem intake of protein has declined in all India level. This is also true for North-eastern region. It may be mentioned here that states which maintained per consumer unit per diem intake of protein above the national average as per report 238/1-11. all of them are now also consuming more than national average of 31.4 gm. In terms of calorie, the people of the North eastern states are not getting, on the average, minimum requirement of 2400 K. cal per capita per day in rural areas. This is, however, also true in case of India as a whole.

Table 15.3 : Consumption of Nutrition Per Consumer Unit Per Day

Sr No.	*State/Union Territory*	*Rural*		*Urban*	
		Protein (gm)	*Calorie (K cal)*	*Protein (gm)*	*Calori (K Cal)*
1.	Delhi	99	3040	78	2572
2.	Himachal P.	98	3190	82	2823
3.	Tripura	80	3027	78	2857
4.	Pondicheri	66	2321	68	2410
5.	Tamil Nadu	66	2394	58	2239
6.	Mysore	77	2839	67	2372
7.	Kerala	50	2023	56	2103
8.	Andhra Pradesh	73	2666	68	2617
9.	Uttar Pradesh	102	3198	84	2616
10.	Punjab	111	3711	85	2806
11.	J&K	100	3490	76	2772
12.	Haryana	115	3652	89	2789
13.	Manipur	79	3109	77	2948
14.	Goa	68	2350	74	2614
15.	Meghalaya	75	2577	73	2623
16.	West Bengal	63	2311	70	2431
17.	Orissa	66	2533	73	2737
18.	Maharastra	78	2567	77	2502
19.	Rajasthan	102	3213	92	3006
20.	Madhya Pradesh	128	3756	84	2850
21.	Gujarat	82	2822	75	2625
22.	Bihar	83	2732	83	2763
23.	Assam	69	2665	71	2616
24.	Chandigarh	-	-	82	2826
25.	Nagaland	-	-	81	2626
	All India	76	2724	75	2539

Soruce: *EPW*, August 25, 1984, p. 1492

(Caloric and Protein content of food items consumed per diem per consumer unit based on NSS).

Report No. 238/1-1

Table 15. 4 : Per Capita and Per Consumer Unit Intake of Claorie Protein and Fat Per Diem In The Rural North-East India And Indian as a Whole.

State/Country	*Per Capita Per Diem Intake of*			*Per Captra unite in take of*		
	Calorie	*Protein*	*Fat*	*Calotie*	*Protein*	*Fat*
	(K. Cal)	*(0.0 Gm)*	*(0.0 Gm)*	*(K.Cal)*	*(0.0 Gm)*	*(0.0 Gm)*
Assam	1983	49.5	21.0	2406	60.1	25.5
Arunachal Pradesh	2130	61.5	16.4	2620	75.7	20.2
Manipur	2157	55.0	12.3	2664	68.0	15.3
Mizoram	2110	59.3	22.5	2576	72.4	27.4
Nagaland	2216	62.0	15.8	2667	74.7	19.0
Meghalaya	1977	50.9	23.4	2430	62.5	28.7
Tripura	1924	51.1	21.7	2350	62.5	26.5
All India Rural)	2153	64.2	60.2	2683	31.4	74.0

Source : *Sarvikshna* Vol. XXI No. 2 Oct. 1997, DIPI of Statistics, Ministery of Planning and Programme implementation, GOI, New Delhi.

The 50th round data also reveal that per capita per diem as well as per consume unit per diem level of intake of energy, protein and fat vary with monthly per capita expenditure class (MPCE) level and was very low for those persons belonging to the lower MPCE classes and quite high for persons belonging to higher MPCE classes.

Many children of North-East India are also suffering from malnutrition.

There are two approaches commonly used to access the nutritional status of children: One based on food intake and another based on anthropometric measurements. The following are the commonly used indicators of undernutrition that are based on anthropometric data: (i) Weight for age (ii) Height for age (iii) Weight for height.[4]

The NRTC (Calcutta. 1988) study shows that in Mizoram on the basis of Rao's ratio of weight-height, 26-28 percent of boys & girls in the age group 1-5 years are having protein energy malnutrition while for the 72-74 percent the nutritional status is normal. An extreme form of malnutrition is Oedema. Any children in the age group 0-14 years, about one percent in Aizawl & little more than 1 percent in Chhimtuipui district were suffering from Oedema. Vitamin A deficiency in eyes was observed in a little over 8 percent children in the state. The Central Goitre Survey team in 1978 carried out a survey on the prevalence of Goitre & found Mizoram as Goiter endemic area.[5]

Factors Affecting Health and Nutrition of the people of North-East.

Good Health of the people is influenced by a complex of social, economic, political and environmental and biological factors. Some of the important factors like (i) Genetic (ii) Climatic (iii) Income of the people (iv) availability of regular food grains and other items of consumption, and the role of the public distribution system. (v) Health education (vi) Development of road and communication facilities (vii) Literacy specially female literacy (viii) Supply of pure drinking water (ix) Social environment (x) Govt. policy that influence health and nutrition are discussed one by one.

The extent and type of available health facilities and utilisation of these will be discussed in the heading. "Health Facilities in the North-Eastern Region".

Genetic

The genetic factors are external factor to every human being and it is very difficult to control and cure certain genetic diseases.

Climate

Climate is one of the important factor that affect people's health. Pollution free water, air has a very positive impact on the health of a person. Except few towns, North-eastern region is free from pollution. Some research work may be undertaken to examine the impact of pollution free climatic condition on our health. It is observed that majority of the expecting mother gives normal birth in the hilly areas than in the cities.

Income of the household

Health depends on income of the household. Generally there is a positive relationship between income & consumption. As majority of the people in North east India are living below poverty line, so they are not getting adequate nutrition. Unemployment problem in the region is also increasing.

Regular Supply of food grains and other items of consumption

In the North-east region more particularly hilly region supply of food grains and other items of consumption is very irregular. For example

Jenging is a place of Arunachal Pradesh. It is about 36 Km from Ying Kong which is the district town of upper Siang district. Since June, 2000, a large number of places like Karko, Gosang. Mayung, Ramsingh, Maising Janboo, Bomdo, Mijjing are communicationally isolated from the rest of the country due to flood in the region. It is reported that many people of these areas are not getting sufficient food. This is because all the roads from Ying Kiong to Mijjing is traceless due to floods. It may be mentioned that in Jenaging telephone facility with (STD) is there. But since June 2000 Telephone service is out of order in that area.[6]

We know that North-eastern region is a food deficit area. Annual shortage of food grains in this area is estimated to be two million tonnes, so, the area depends on other parts of the country for import of food grains. Public distribution system can play an important role in this area. But due to lack of road and transport facilities public distribution system become inoperative in this area at least half of the year (more particularly in hill areas). Road and transport system is disrupted in most part of the region during rainy season due to land sliding, floods, damage of bridges etc.

Health education

Most of the people more particularly in the rural areas do not have any knowledge of keeping good health. It is found that about 60 per cent of the people of the tribal areas of Barak Valley do not drink boiled water. At the time of disease they generally call Hojai (priest) and sacrifice Pigs. Hens etc. In Arunachal Pradesh whenever any people becomes ill the poor persons generally sacrifice Pig and the rich people sacrifice Mithun, the cost of one is approximately Rs.10,000.

Literacy

Health and nutritional condition depends on the literacy rate more particularly female literacy. Table- shows that female literacy rate in all the states of North-east except Arunachal Pradesh is above 3 the National rate, but it is much lower than Kerala which is demographical one of the developed states. Mizoram ranks two in female literacy rate in India in 1991. By raising female literacy rate it is possible to improve the health and nutrition of the people. There are many households where there are no literate member in the family. For example in 1981 in Arunachal Pradesh 54.91 per cent of the rural households did not have any literate member in their family. The position of the other states was Meghalaya 43.94 per cent, Manipur 23.43 per cent, Nagaland 29.11 per cent, Mizoram 11.93 per cent.

Supply of Pure Drinking Water

WHO'S statistics shows that 80 per cent of diseases are due to inappropriate use of water and poor environmental and sanitation facilities which includes lack of disposal facilities of human waste and solid materials. Cholera, dysentry, gestro-enterties, typhoid are common epedemics spread through water. But in the majority of the areas, people are not getting pure water. For example in Assam 45.86 per cent of the household is getting safe drinking water facilities. The condition of the other states of the North-eastern region are not so bright.[8] There are many habitats in the North-eastern region which is facing water quality problem. Supply of water in most of the villages are also not regular, Statistics available from Silchar Civil Hospital reveals that out of 42282 cases reported in the OPD Department 19045 people were the victims of Diarrhoea, 8543 persons were the victims of Dysentery in 1994.[10]

Developments of Roads and Communication

A developed road and communication system of a region lowers the incidence of mortality rate. People can avail of the better treatment facilities from the nearby hospitals and dispensaries of the town. But many areas of North-eastern region become isolated from the rest of the country during rainy season. In many hilly areas except one bus in a day there is no other transport facilities.

North eastern region is very backward from the stand point of transport. In all the North eastern states except Assam and Tripura, road per 100 sq. km is lower than all India average of 73 km. The percentage of surface total road length is lower than national average of 55.5km in all the North Eastern states. However, in terms of road per thousand population, the position of all the north eastern states is better than all India average of 2.85 km (Table 15.5). Statewise Registration of Motor vehicles is shown in Table 15.6.

Table 15.5 : Road Length in NE Region

(As on 31-3-96)

State	*Total Length*	*Surfaced*	*Unsurface*	*Road per 100 sq.km.*	*Road per '000 popu-lation*	*% of surface Road to total road length*
1	*2*	*3*	*4*	*5*	*6*	*7*
Arunachal P.	10240	3692	6548	12.2	11.38	36.05
Assam	68079	11421	56658	86.8	3.05	16.78
Manipur	10760	3595	7165	48.2	5.98	33.41
Meghalaya	8391	3946	4445	37.4	4.66	47.03
Mizoram	6910	2360	4550	32.8	9.87	34.15
Nagaland	13732	2515	11217	82.8	11.44	18.31
Tripura	14726	4577	10149	140.4	5.45	31.03
Total	132838	32106	100732	52.8	4.23	24.17
All India	2403634	1334078	1069556	73.0	2.85	55.50

Source: Basic statistics of NE Region, 2000, p. 110

The railway network in the north eastern region is also very poor. Till now Meghalaya is not included in the railway map. The total gauge in Arunachal Pradesh, Manipur and Mizoram is also very negligible i.e. less than 2 kms in these states (Table 15.7).

Social Environment

In addition to the above mentioned factors society's attitude regarding consumption of harmful things like Ghaga, Charos etc. is also an important factor that determine health of the people. By creating social awareness regarding AIDS, drug addiction, prohibition, the health of the society can also be improved.

Government policy

It is also possible to improve the health and nutritional status of the people if the poor families are given subsidised food, treatment and medicine. Again by publicity and health education the health consciousness of the people may be created.

Health Facilities in the North East

The health facilities of North-East is very poor. Most of the village and hilly places do not have any roads. In most of the hospital

Table 15.6 : Statewise Registration of Motor Vehicles *(As on 31-3-97)*

State	*All Vehicles*	*Trucks*	*Buses*	*Jeeps*	*Cars*	*Taxis*	*Two Wheelers*	*Three Wheelers*	*Tractors*	*Trailor*	*Others*
1	2	3	4	5	6	7	8	9	10	11	12
Arunachal P.	2114	2878	665	2260	2340	299	10605	1430	333	155	179
Assam	378111	60386	10293	11414	41411	6825	297367	10207	7251	5758	16199
Manipur	65275	5282	1742	4821	3525	303	46229	1987	726	499	159
Meghalalya	47995	10570	1955	7408	8274	2801	13703	399	414	2070	401
M:zoram	20310	2921	634	5240	1668	1600	7910	-	163	167	16
Nagaland	107560	25507	3302	21925	21268	2121	21786	7163	2249	677	1562
Tripura	36767	4949	1320	3432	3742	526	19056	2231	108	986	417

Source: Basic Statistics of North Eastern Region, 2000, NEC, Shillong, p. 113.

Table 15.7 : Railway Network in NER *(As on 1998-99)*

State	*Broad Gauge (BG)*	*Metre Gauge (MG)*	*Narrow Gauge (NG)*	*Total Gauge*
Arunachal P	Nil	126	Nil	1.26
Assam	902.94	1470.58	Nil	2373.52
Manipur	Nil	1.35	Nil	1.35
Meghalaya	-	-	Nil	-
Mizoram	Nil	1.50	Nil	1.5
Nagaland	763	5.22	Nil	12.85
Tripura	Nil	44.72	Nil	44.72
Total	910.57	1524.63	Nil	2435.20

Source : Basic statistics of NE Region, 2000, p. 111.

the post of doctors remain vacant year after year. In rural areas/ tribal areas the poorer members of the societies rarely call in doctor in their house. They call in medicine seller. In the tribal villages there are two types of medicine sellers - (i) registered & (ii) unregistered. The visiting charge of the registered medicine sellers is a bit higher than that of an unregistered medicine seller. They rarely call a doctor because the visiting charge of the doctor is very high. The visiting charge of a doctor in the tribal areas of Barak Valley ranges from Rs.10.00 to Rs.200.00 (in 1984) depending on the distance travelled by him and gravity of the case.[11]

Number of doctors and doctor population ratio is very low in this region when we compare these figures with India as a whole (Table 15.8) . Population serve per bed in Assam (1968) and Tripura (1647) are higher than India as a whole (1324) (Table 15.9). No. of primary Health Centres and Sub-centres is also very low (Table -15.10). Moreover in certain indicators of health services like population per bed have decreased during 1985-1992 period. In 1985, population per bed was 1441 but in 1992 it became 1529. During the same period growth rate of bed in all India level was 19.85 per cent; but in the North-Eastern Region, it was only 163 and Meghalaya by 430 during 1985-92 period. However. doctor population ratio improves during the same period. it was 1:16028 in 1985 and 16412 in 1992.

Table 15.8 : Statewise Number of doctors and doctor Population Ratio in North Eastern Region - 1991

State	*Doctors engaged under Govt. Agency*	*Population served per Doctor*	*Period to which data relates*
1	*2*	*3*	*4*
Arunachal Pradesh	233	1:3536	31:12.90
Assam	2660	1:8750	31.12.91
Manipur	684	1:2675	31.12.91
Meghlaya	322	1:5357	31.12.90
Mizoram	146	1:5000	31.12.91
Nagaland	202	1:5401	31.12.91
Tripura	673	1:3822	31.12.90
All India	39466	—	—

Source : Basic Statistics of NE Region, 2000, p. 117

Table 15.9 : Number of Hospitals and Beds According to Ownership *(As on 1.1.1992)*

State	*Government*		*Local Bodies*		*Pvt. & Vol. Org.*		*Total*		*Population*
	Hosp.	*Beds*	*Hosp*	*Beds*	*Hosp*	*Beds*	*Hosp*	*Beds*	*Served Per Bed.*
1	*2*	*3*	*4*	*5*	*6*	*7*	*8*	*9*	*10*
Arunachal Pradesh (1.1.91)	18	1091	—	—	—	—	18	1091	755
Assam (1.1.91)	141	9687	47	982	80	1992	268	12661	1968
Manipur (1.1.91)	21	1360	—	—	4	100	25	1460	1203
Meghalaya	4	985	—	—	4	650	8	1635	1055
Mizoram	11	818	—	—	3	310	14	1128	627
Nagaland (1.1.91)	31	1114	—	—	—	—	31	1114	979
Tripura	23	1531	—	—	—	—	23	1531	1647
All India	4411	411868	346	23347	6417	206888	11174	642103	1324

Source : Basic Statistics of North Eastern Region, 1995, NEC, Shillong, p. 130.

Table 15.10 : Number of Sub-Centres, PHCs and ClCs Functioning

(As on 31.12.97)

SLNo	*State/UT*	*Sub-Centres*	*PHCs*	*CHCs*	*Date of Latest Report*
1	*2*	*3*	*4*	*5*	*6*
1	Arunachal Pradesh	245	47	9	31.3.97
2	Assam	5280	619	105	31.3.95
3	Manipur	420	72	16	31.12.97
4	Meghalaya	377	82	13	31.12.97
5	Mizoram	324	38	6	30.9.97
6	Nagaland	244	33	5	31.12.97
7	Tripura	537	56	11	31.12.97
8	All India	136815	22962	2708	

Source: Basic statics of NE Region, 2000, p. 119.

However, per capita expenditure on hospitals and dispensaries is the highest in Arunachal Pradesh i.e. Rs.174.26 among all the states of India. In Kerala that figure is Rs. 51.44. The position of other North-Eastern states is as follows:[12]

Assam Rs.7.39, Mizoram - Rs.56.83, Meghalaya Rs.61.89, Nagaland - Rs.26.18, Tripura - Rs.42.41 and All India - Re.21.91.

Similarly in terms of bed per 1,00,000 population, except Mizoram and Arunachal Pradesh the figure in all states of North-Eastern region is less than 100.

In Kerala that figure is 263-2. The position of Assam is very bad if we consider dispensaries or hospitals or bed per 1,00,000 population.

Remedial Measures

From the foregoing analysis it becomes clear that the health and nutritional level of the people of northeastern states do not much differ from the people of the nation as a whole. The environment of this region is much better than many other parts of India because there is comparatively less air and water population. But health services in this region is not satisfactory. Owing to various reasons, health worker seldom visit the villages. It is confirmed by NHFS 1992 that visit of the health worker in the rural areas either for antenatal care or for Family Planning follow up services is very low. The rates (in percent) in different North-eastern states go like this. Arunachal Pradesh-2.5, Manipur-3.5, Meghalalaya-4.3., Mizoram-

16.8, Nagaland-5.6, Tripura-5.5. Assam-7.9, against all India average of 28.5.[13] The percentage of children 1-47 months who are underweight (below 2 SD of the WHO/NCHS Standard) for their reported ages is the highest in Assam among the north-eastern states.[14] Moreover, many people cannot bear the cost of medicine because now-a-days Allopathic treatment is very costly for the poor people. So in order to improve the health condition of the people the following suggestions may be given:

(i) Provision for low cost medicine more particular Homeopathic and Ayurvedic may be made.
(ii) Research on indigenous medicine/medicinal herb should be made.
(iii) Effort may be taken to make environment pollution.
(iv) Effort should be taken to raise female literacy rate.
(v) Health education may be given to the households by organising seminars, fairs, posters and also by community meeting.
(vi) Road and transport system should be developed.
(vii) Through revamped public distribution system the insecurity of food may be removed.

NOTES

1. Statistical Outline of India 1998-99, Tata Services Ltd., Dept. of Economics and Statistics, Mumbai. p-46 & Dhingra, I.C.: *The Indian Economy*, Sultanchand and Sons, New Delhi, p.93,
2. *EPW*, Dec., 25.1999, P. 3632.
3. *Jusgasanka*, a Bengali daily News-paper published from Silchar, Dated 16th November, 2000.
4. Geeta, S and Swaminathan, M: Nutritional Status of Slum Children of Mumbai: A Socio Economic Survey; *EPW*, April 6,1996, p.896.
5. Guha Thakurta, S.N.: Developmental strategy and poverty Alleviation Programme in the book Datta Ray, B. and others (ed) *Population Poverty and Environment in North-East India*, p. 278.
6. Collected from Sri Arun Choudhary of Jengging, Arunachal Pradesh.
7. NCERT: University Primary Education of Rural Girls-in India A, June 25. 19912.P.168.
8. For details, please concern Census of India, occasional paper no.15 of 1994. Housing and Amenities.
9. Goswami, Pranay Jyoti : Status of Potable Water Supply in the Rural areas of North East India with special reference to Barak Valley in the book Datta. Ray, B & Athparia, R.P.(ed) Water and water resource management, Omson Publication. New Delhi-1999.
10. Assam University; Silchar: Silchar-Problems of Urban Development of Growing City, p.58.

11. Goswami, Pranay Jyoti: Tribal in Transition - A Socio-Economic Study, Jawahar Publisher & Distributors. New Delhi- 995, p.87-89.
12. *EPW*. April 22, 1995, p.904.
13. Bhat, P.N. Mari & Zavier Francis: Findings of National Family Health Survey Regional Analysis. *EPW* 2000. p.3008.
14. *Ibid*., p.3028.

16

TRIBAL POPULATION OF THE BRAHMAPUTRA VALLEY UNDER CHANGING SOCIO-POLITICAL SITUATION

— *N.N. Bhattacharyya*

A tribe is a loosely organized group of primitive people. Generally such a racial group is united by language and customs, living as a community under one or more chiefs. The tribes are generally classified on the basis of territory, language, occupation, physical characteristics and cultural contracts. Some Indian social scientists have accepted certain overtones or indicators of tribals of India and accordingly they are non-Aryan particularly of pre-Aryan stocks. Indian tribals are small groups who live in forests, hills and some isolated plains away from the non-tribal majority. They are followers of some animistic religion and are not adherent to any great religions. They are primitive and socio-culturally backward. They speak none of the major Indian languages and each tribe has its own undeveloped language or dialect. Their economy is based on hunting, food gathering and on shifting or primitive cultivation. to a demographer, a tribe lives in a small settlement with low density of population, males outnumbering females in the middle age-group and their average longevity is low. By these indicators it is very difficult to define a tribe of Assam. The tribes of Assam are not primitive, nor they are very backward and all of them are not hill-dwellers or forest-dwellers. They are hardly cut-off physically and socio-culturally from the non-tribals, (Bhattacharyya, 1990). However, under Articles 341 and 342 of the Constitution of India, the President of India specified the scheduled castes and scheduled tribes. The schedules castes, because of certain traditional inabilities and scheduled Tribe owing to their geographical isolation, have not been able to derive in full measures the benefits of the economic development of the country. That is why the constitution of India has prescribed certain protective measures and safeguards for the communities. Thus the tribals have drawn special attention of the government of India since independence. In order to solve their problems quicker and in consolidated manner, states like Nagaland, Meghalaya, Tripura, Manipur, Mizoram and Arunachal Pradesh have been created and recently Uttarakhand, Bananchal and Chhatisgarh are in the process of formation. Largest proportion of the total population of scheduled tribe is in Lakhadweep (93.82 per cent) and Mizoram (93.55 per cent), followed by Nagaland (83.99 per cent) and Meghalaya (80.58 per cent). A large number is scheduled tribes are found to live in Madhya Pradesh, Bihar, Uttar Pradesh, Orrissa, Assam etc. In Assam less productive and forested lands are occupied by the scheduled tribes like Bodo, Garo, Rabha, Tiwa, Karbi, Dimasa, Kachari, Barman, Kuki, Mizo etc. Besides, flood plains of Upper Assam are generally occupied by the Mishing, Deuri and Chutiya tribes. As per 1991 census out of the total population of 22414327 as many as 2874441 persons or 12.82 per cent were scheduled tribes in Assam, they mostly belong to the Mongoloid stocks. They are socio-economically underdeveloped as compared to the non tribal communities. They even feel socio-economically and politically suppressed and exploited by the non-tribals.

Since the starting of this century, following large scale influx of immigrants especially from undivided Bengal and Nepal into Assam, a sense of fear-psychosis of insecurity has developed among the indigenous people including tribals. Thus the tribals being alarmed at this have ethinically united to form a tribal league in Assam as early as 1933. It is said that this organisation succeeded in fulfilling certain grievances of the tribal people like secured political representation in the Legislative Assembly, formation of tribal belts and blocks. After independence also the people of Assam including non-tribals have developed a feeling against the central government that they are neglected and exploited. They have been passing days in anxiety for safeguard of socio-economic and political identity and future security within the Indian nationhood. Recently most of the tribes of Assam have been demanding separate homelands for them like Bodoland. Autonomous state for Karbi and Dimasa groups, autonomous districts for Rabha-Hasong, Mishings and Tiwas. In the light of the above discussion an attempt have been made here to study the spatial distribution pattern of the different tribal groups and the locational attributes contributing to their territorial isolation, socio-economic and political demands of the tribals and the prospect of a viable solution to their problems in the Brahmaputtra Valley.

The total population of the Brahmaputra Valley as per 1901 census was 26,19,077 persons and the tribal population in that year was 3,69,252 persons, constituted 14.1 per cent of the total population of the valley. The proportion of the tribal population of the valley decreased to 10.67 per cent in the year 1971 and the same was computed to 12.71 per cent in 1991. The fall of the proportion of the tribal population to the total population of the valley during the period 1901-1991 may be attributed to the very high growth of population of the Brahmaputra Valley as a result of huge influx, of migrants into the valley from outside the region and also the natural growth of all the indigenous groups of population including non tribals. The following Tables16.1 and 16.2 shows the districtwise populations of the tribal population and their concentration in the Brahmaputra Valley in 1901 and 1991. The tribal population of the Brahmaputra Valley has been increasing in all the districts during 1901-1991 but their spatial concentration is more significant in the undivided Lakhimpur district which includes present districts of Dhemaji, Lakhimpur, Dibrugarh And Tinsukia district. It is also observed that the tribal people generally avoid strong urban influence and their, spatial concentration have a direct co-relation with the distance from the large towns. Most of the tribal people seem to be allergic to Urban influence and spatially they are found to be distributed in areas beyond the zone of Urban influence. Dhemaji is the least Urbanized district of the Brahmaputra Valley with more concentration of tribal population. Besides the influx of migrants is a bit lower in Upper Assam districts than in the lower and middle Assam Valley districts. The location quotients are higher in the undivided districts of Lakhimpur, Kamrup and Goalpara, normal in Darrang district and lower in Nowgaon and Sibsagar District.

The hilly areas of Assam and the forested regions of the Brahmaputra Valley used to be the areas of sparse population or India's lebensram. These areas were somewhat unfavorable for settled cultivation and only backward tribal people occupied such areas. A few decades ago such areas were extensive and at present these areas were swamped the waves of immigrants from the overpopulated parts of Assam. Other states of India, Bangladesh and Nepal. This process has resulted in deforestation and transformation of the landscape which has become unconducive for tribal environment. Immediately after partition of the country a major chunk of the post partition refugees from Bengal was rehabilitated by the government in the tribal areas of the Brahmaputra Valley. Thus they were victims of institutional mechanisms and so the tribals of the region have been loosing faith of the different social and economic institutional arrangements of the government. To them all these institutional mechanisms are anti-tribal. It is to be noted that the shift in the main criteria of the Indian states reorganization from the language to ethnic and cultural base in the North East India Reorganisation Act 1972 and the formation of a number of small economically unviable states for the hill tribes have provided incentives to the aggrieved plain tribals of the Brahmeputra Valley to raise the demands of separate states (Barthakur. 1995).

Table 16.1 : Districtwise (Undivided) tribal population in the Brahmaputra Valley in 1901 and 1991.

1901 Districts	*1999 Present Districts*	*1901 Total Population*	*1901 Total Tribal population*	*1991 Total popu population*	*1991 Total Tribal popri.*
1. Goalpara	Kokrajhar Dhubri Bongaigaon & Goalpara	462052	104608	3590806	618362
2. Kamrup	Barpata Nalbari & Kamrup	589187	79782	4382985	494333
3. Darrang	Darrang & Sonitpur	337313	61969	2705117	387455
4. Lakhimpur	Lakhimpur Dhemaji Dibrugarh & Tinsukia	371396	26203	3212124	521881
5. Nowgaon	Nagaon & Marigaon	261160	54521	2532463	168331
6. Sibsagar	Sibsagar Jorhat & Golaghat	597969	42169	2565297	224719

Table 16.2 : Districtwise (Undivided) Concentration of the Tribal Population in the Brahmaputra Valley in 1901 and 1991.

1901 Districts	*Percentage of Tribal population to the total population of the district.*		*Location Quotient (L.G) for tribal population.*	
	1901	1991	1901	1991
1. Goalpara	22.64	17.22	1.61	2.19
2. Kamrup	13.54	11.51	0.96	2.88
3. Darrang	18.37	13.95	1.30	1.93
4. Lakhimpur	7.06	16.19	0.50	5.89
5. Nowgang	20.88	6.65	1.48	0.52
6. Sibsagar	7.05	8.76	0.50	1.08

* The concentration of tribal population has been shown by Location quotient.

$$LQ = Pt/PT$$

Where LQ = Location Quotient, Pt = Percentage of the population group T to the total population of the district.
PT = percentage of the population group T to the total population of the region.

In the past, the economy of tribal areas of Assam was in a state of complete self-sufficiency, having developed their, own way of life without any help from outside. At that time the tribes of Assam enjoyed the advantage of a well balanced ecology with the available natural resources including foodstuffs. They had a non-exploitative social structure, freedom from indebtedness and other forms of dependence on non-tribals. But now their traditional style of life has been disrupted by disastrous effect of Western culture and materially by non-tribals who invaded their habitat. After independence when the Government of India introduced some developmental programmes for promoting social and economic well being of the people of the tribal areas, the subsistence structure of the tribal economy broke down. The constitution of the tribal belts and blocks in different tribal areas also failed to achieve basic objectives of protecting the tribal lands from being transferred to the non-tribals and immigrants owing to the institutional and administrative lapses. Therefore, land reforms should be implemented in such a way so that the tenant cultivators become the landowners and landless agricultural labourers could acquire proprietary rights in agricultural lands in the tribal areas of the Brahmaputra Valley. Similarly medical services, literacy electricity and other infra-structural facilities should be extended to all tribal areas of the valley for economic, social and cultural progress. Unless such measures of some other appropriate and planned measures are taken at this stages of socio-economic transition of the Brahmputra Valley, the whole state will remain under developed in 21st Century and too many demands for separate tribal states will be raised by the different political and non-political tribal organizations of Assam.

REFERENCES

Bhattacharyya, N.N., "Linguistic and Tribal Regions of North-East India" in Mandal, R.B.ed., *Patterns of Regional Geography : An International Perspective*. Vol.2 (Concept Publishing Company, New Delhi, 1990, p.151).

Barthakur, P., "The Bodo Tribal in the Changing Socio-Political Landscape of the Brahmaputra valley : A study in Political Geography" Unpublished Ph.D. Thesis, Gauhati University., 1995, p.223.

Jones, E., *Readings in Social Geography*, London, 1975.

Mukherji, A. B., *The Chamars of Uttar Pradesh; A Study in Social Geography*, Inter-India Publications, Delhi, 1980.

17

JUDICIAL INTERVENTION IN THE MANAGEMENT OF FORESTS IN THE NORTH EAST INDIA

— *S. K. Agnihotri*

Forests play a vital role in the life of the people of the northeast India comprising the states of Arunachal Pradesh, Assam, Manipur, Meghalaya, Mizoram, Nagaland and Tripura. The National Forest Policy of 1988 has recognised their role in maintaining sustainable ecological balance, environmental stability, bio-diversity conservation and food security for the entire country. The policy aims to have a minimum of one third of geographic area under forest cover. However for the hill areas, the policy enjoins maintaining two thirds of the area under forest cover to prevent erosion and land degradation and to ensure the stability of the fragile eco-systems.

The management and exploitation of forests in the north east India has always not been as per the national policy or in the long-term interest of the people. Indiscriminate felling and unscientific management has resulted in the depletion, degradation and endangering natural regeneration of the forests. Since the state government agencies concerned with the forest management and conservation did not take adequate action, the Supreme Court had to intervene in 1996 to direct the states to ensure:

1. protection of adequate forest cover;
2. scientific management of forests, and
3. establishment of appropriate authorities to implement its directions.

This paper discusses the state of forests in the seven states of the north east with separate data for hill and tribal districts and initiatives taken by the Supreme Court to ensure sustainable development of forests.

The demand for wood as timber and fuel has reached a level in India which exceeds the capacity of the forest to meet requirements and its important contribution to the well being of rural communities can no longer be taken for granted. A detailed study of the overall trends in various regions provides an opportunity to identify developments and tendencies which might be dangerous if their implications are not immediately recognized and appropriate corrective action taken to avert harmful consequences. Though the north-eastern region is capable of increasing its wood supplies by scientific management of forests but this capability is being curbed by conversion of forest to agricultural uses and destruction of forest by shifting cultivation and destructive logging.

For the vast majority of people in the region, consumption of manufactured wood products is trivial and the most important forest product for them is fuel wood. If greater supplies are available to rural communities, wood as a source of energy and structural material will have substantial potential to improve their well being. It has the advantage that it can be produced by the efforts of the local people with the minimum of external resources in transport and manufacture. Where fuel wood is in short supply its collection often results in the over exploitation of' the forest resource, causing destruction of the source of future supplies and destroying the forests ability to perform its conservational role in protecting watersheds and preventing soil erosion and flooding. Knowledge about the value of wood to rural communities and possible methods and costs involved in arranging the supply in ways in which there will be no harm to the forest and environment is inadequate. Therefore formulation of an adequate policy in the area is of immediate necessity. It is extremely important to find solutions capable of being adopted by the rural communities.

Economic Background

In order to indicate the role and potential role of the forestry and forest industries sector in contributing to the economy of the region it is necessary to outline the major economic characteristics and problems of the region.

There are large areas of land in mountain regions of little value for settlement. Similarly, in the plains, a considerable area is floods prone. A great part of the population of the region is rural. There is low rate of employment due to inadequate employment opportunities. There is considerable under employment and a rapidly growing labour force. The income levels are generally low. A very large number of people have little involvement in commercial activity and such involvement is largely to meet subsistence requirements for food, clothing and articles for domestic consumption. The extent of employment in industry is limited. There are lower growth rates, low rate of investment and low rates of growth in the industrial sector.

Forest Products in the Economy of the Region

Timber contributes to the economics of the states in the region in two main ways: first as a source of energy and raw material in the domestic economy, particularly of the rural community, and secondly as a source of earning by way of export to other states outside the region. For a vast majority of the population of region wood makes its contribution to their well being in the form of fuel wood and un-processed timber. The average apparent domestic consumption of sawn wood is very low. Panel products play a very minor role in the consumption of the states in the region. Similarly, consumption of paper is at a very low level. The export of forest products contributed an important component in the economy of some states till the orders were passed by the Supreme Court in December 1996.

Other Social and Economic Functions of the Forest

The forest is involved in the social and economic system of the region in many other ways not fully represented by statistical and economic indicators. Forest play an important environmental role in protecting watersheds, conserving water resources and providing areas suitable for recreation and for the protection of wild life. Forest also contribute to the welfare of rural communities through their supply of other minor products such as forage for animals, fruits, resins, gums and dyestuffs. Of course the forests directly contribute to the well being of the community of the region through their material production, the wood for fuel used in cooking and heating, the structural material of sawnwood and panel products for housing, construction, furniture, wooden goods and packaging and the paper used in education and communications and a wide range of functions essential to the efficient functioning of the community.

Forestry production makes use of land with lower potential for agricultural production and contributes to the employment of otherwise underemployed rural population. Forest production may be carried out and jointly with agriculture through the use of *toungya* system and forest grazing.

The forest is regarded as the automatic source of land for agriculture with the intense pressure of increased food production. Haphazard destruction of forest is not only wasteful of existing wood and potential forest production but is also profligate in the use of the land, leading to soil erosion, deterioration of water supply and flooding, and frequently resulting in the land being subsequently abandoned as wasteland. The harmful effects of flooding and siltation arise often at great distances from the location where the forests have been destroyed.

Spread of Forestry Management

In the present context forestry profession has a wider clientele viz.

(1) farmers wanting to include trees in their agriculture systems,
(2) villagers wanting wood lots for fuel and other materials;
(3) recreationists wanting places to relax and play;
(4) tribal people who traditionally have had access to and control over local forest resources until population pressures began to cause competition for those resources;
(5) land less people wanting food, fuel fibre and other forest produce that are available to them only on public lands; and
(6) conservationists wanting to protect the diverse resources and environments

This widened clientele has changed the concept of forestry management and its main areas of focus could be

(a) managing remaining natural forest areas in a sustainable way;
(b) involving local communities in the protection and management of remaining natural forest resources;
(c) identifying critical watersheds and undertaking programme of conservation;
(d) finding appropriate solutions to minimize forest degradation by shifting cultivation;

(e) establishing networks of protected areas to conserve bio-diversity;
(f) strengthening conservation, collection and storage of germplasm together with tree improvement to ensure productivity gains in reforestation programmes;
(g) utilisation of non wood forest products and plantation grown wood;
(h) promoting community forests and agro forestry;
(i) promoting amenity forestry; and
(j) protection and welfare of wild life

This trend should be taken into account while undertaking educational preparation of future forestry professionals.

Decline of forests

Decline of forest has occurred in the region both due to non-anthropogenic and authropogenic factors. Non-anthropogenic factors are those which have occurred in the absence of human influences. These include :

(i) low species diversity;
(ii) formation of even age stands (cohorts) on marginal sites, usually following a natural catastrophic event such as wild fire or high winds, which has destroyed the original vegetation;
(iii) development of chort senescence;
(iv) a triggering mechanism such as defoliation, drought, excess rainfall or high winds; and
(v) invasion of declining trees by secondary pathogenes or insects.

The main anthropogenic factors involving human activity are :

(a) atmospheric deposition of pollutants;
(b) use of forests and woodlands for pastroral purposes;
(c) improper harvesting of forest products;
(d) single species plantation;
(e) accidental introduction of pest species;
(f) fire exclusion; and
(g) altered hydrologic systems

Besides these two factors, climate change also influences the existence, abundance and distribution of forests.

The decline of forests can be managed by diagnosis of decline and monitoring and assessment. The management system recommended to reduce losses will include the following:

(1) avoidance of overgrazing and associated soil compaction in forests which are prone to decline events;
(2) selection of species varieties and geographical areas which are adopted to sites being planted while establishing plantations.
(3) surveillance at entry point to intercept potentially damaging insects, fungi etc;
(4) use of prescribed fire to help keep the desired vegetation in a healthy condition;
(5) liming and fertilization in case the forest decline is associated with soil nutrient deficiency; and
(6) comperhensive analysis prior to implementation of development projects where changes in the pattern on volume of water flow in rivers and estuaries are anticipated;

Forest Resources of the Region

Arunacal Pradesh

The state has got recorded forest area of 51,540 sq.km. out of the total geographic area of 83,743 sq.km. However, as per the State of Forest Report, 1997 the actual forest cover of the state is 68,602 sq.km. which constitutes 81.90 percent of the geographic area. There is a net decrease of 19 sq.km. in forest cover in 1997 assessment compared to 1995 assessment. Table 17.1 shows forest cover change matrix of the state of Arunachal Pradesh.

Assam

The state has got recorded forest area of 30,708 sq.km. out of the total geographic area of 78,438 sq.km. However as per the State of Forest Report, 1997 the actual forest cover of the sate is 23,824 sq.km. which constitutes 30.40 per cent of the geographic area. Table 17.2 shows forest cover change matrix of the state of Assam.

Table 17.1 Forest cover change matrix of Arunanchal Pradesh *(sq.km)*

1995 (Data. Dec.91 & Nov.92)	*Assessment 1997 Assessment (Data- Nov.94-Nov.95)* Dense forest	Open forest	Mangrove	Scurb	Non-forcst	*Total 1995*
Dense forest	54,030	126	—	—	20	54,176
Open forest	125	14,265	—	9	46	14.445
Mangrove	—	—	—	—	—	
Scurb		24	—	330	—	354
Non-forest		32	—	76	14,660	14.768
Total 1997	54,155	14,447	—	415	14.726	83,743
Net change	-21	2		+61	-42	

(*Source*: State of Forest Report, 1997)

Table 17.2: Forest cover change matrix of Assam *(sq.km)*

1995 Assessment 1997 (Data- Dec.91 & Nov.92)	*Assessment (Data- Dec.93, April -Nov.94)* Dense forest	Open forest	Mangrove	Scrub	Non-forest	*Total 1995*
Dense forest	15,489	73	—	1	131	15,694
Open forest	59	8,024	—	8	276	8,367
Mangrove	—	—	—	—	—	—
Scrub		5	—	599	19	623
Non- forest	—	174	—	27	53,553	53.754
Total 1997	15,548	8,276	—	635	53,979	78,438
	-146	-91	—	+12	+225	

Source : State of Forest Report, 1997

Manipur

The state has got recorded forest area of 15,154 sq.km. out of the total geographic area of 22,327 sq.km. However as per the State of Forest Report, 1997 the actual forest cover of the state is 17,418 sq.km. which constitutes 78.00 percent of the geographic area. Table 17.3 shows the forest cover change matrix of the state of Manipur.

Meghalaya

The state has got recorded forest area of 9496 sq.km. out of the total geographic area of 22,429 sq.km. However as per the State of Forest Report, 1997 the actual forest cover of the state is 15,657

Table 17.3 : Forest cover change matrix of Manipur (sq.km)

1995 Assessment (Data. Dec.91 & Nov.92)	*1997 Assessment (Data Oct.94-Nov.94)* Dense forest	Open forest	Mangrove	Scrub	Non-forest	*Total 1995*
Dense forest	4,571	693	—	—	54	5,318
Open forest	366	11,325	—	20	529	12,240
Mangrove	—	—	—	—	—	—
Scrub	—	237	—	891	276	1,404
Non-forest	—	226	—		3,139	3,365
Total 1997	4,937	12,481	—	911	3,998	22,327
Net change	-381	+241		-493	+633	

Source : State of Forest Report, 1997

sq.km. which constitutes 69.80 per cent of the geographic area. Table 17.4 shows the forest cover change matrix of the state of Meghalaya.

Table 17.4: Forest cover change matrix of Meghalaya (sq.km)

1995 Assessment (Data. Dec.91 & Nov.92)	*1997 Assessment (Data- Dec.93)* ense	Open forest	Mangrove	Scrub forest	Non-forest	*Total 1995*
Dense forest	4,043	—		—	2	4,045
Open forest	1	—		11,593	75	11,669
Mangrove	—	—		—		
Scrub		813		—	3	816
Non-forest	—	20		36	5,843	5,899
Total 1997	4,044	11,613		849	5,923	22,429
Netchange	-1	-56		+33	+24	

Source : State of Forest Report, 1997

Mizoram

The state has got recorded forest area of 15,935 sq.km. out of the total geographic area of 21,081 sq.km. However as per the State of Forest Report, 1997 the actual forest cover of the state is 18775 sq.km. which constitutes 89.10 percent of the geographic area. Table 17.5 shows the forest cover change matrix of the state of Mizoram.

Table 17.5 : Forest cover change matrix of Mizoram *(sq.km)*

1995 Assessment (Data. Dec.91 & Nov.92)	*1997 Assessment (Data- Dec.93 to Feb.94)*					*Total 1995*
	Dense forest	*Open forest*	*Mangrove*	*Scrub*	*Non-forest*	
Dense forest	3,965	313	—	—	3	4,281
Open forest	374	13,632	—	96	193	14,295
Mangrove	—	—	—	—	—	—
Scrub	6	226	—	835	3	1,070
Non-forest	3	256	—	6	1,170	1,435
Total 1997	4,348	14,427	—	937	1,369	21,081
Net change	+67	+132	—	-133	-66	

Source : State of Forest Report,1997

Nagaland

The state has got recorded forest area of 8,629 sq.km. out of the total geographic area of 16,579 sq.km. However as per the State of Forest Report, 1997 the actual forest cover of the state is 14,221 sq.km. which constitutes 85.80 percent of the geographic area. Table 17.6 shows the forest cover change matrix of the state of Nagaland.

Table 17.6: Forest cover change matrix of Nagaland *(sq.km)*

1995 Assessment (Data. Dec.91 & Nov.92)	*1997 Assessment (Data- Dec.93 to Oct.94)*					*Total 1995*
	Dense forest	*Open forest*	*Mangrove*	*Scrub*	*Non-forest*	
Dense forest	3,487	—	—	—	—	3,487
Open forest	—	10,231	—	--	573	10,804
Mangrove	—	—	—	—	—	—
Scrub	—	—	—	64	56	120
Non-forest	—	503	—	6	1,659	2,168
Total 1997	3,487	10,734	—	70	2,288	16,579
Net change		-70	—	-50	+120	

Source : State of Forest Report, 1997

Tripura

The state has got recorded forest area of 6,292 sq.km. out of the total geographic area of 10,486 sq.km. However, as per the State of Forest Report, 1997, the actual forest cover of the state is 5,546 sq. km. which constitutes 52.90 percent of the geographic area. Table 17.7 shows the forest cover change matrix of the state of Tripura.

Table 17.7 : Forest cover change matrix of Tripura *(sq.km)*

1995 Assessment (Data. Feb.92)	*1997 Assessment(Data- Dec.93)*					*Total*
	Dense forest	*Open forest*	*Mangrove*	*Scrub forest*	*Non-*	*1995*
Dense forest	1,819	—	—	—		1,819
Open forest	—	3,716	—	3		3,719
Mangrove	—	—	—	—	—	—
Scrub	—	—	—	138	—	138
Non-forest	—	I I	—	12	4,787	4,810
Total 1997	1,819	3,727	—	150	4,790	10,486
Net change	—	+8	—	+12		

Source : State of Forest Report : 1997)

The forest cover Status of all the seven states of the north east region is give in table 17.8.

Table 17.8 : Loss/Gain in forest cover in north eastern states *(sq.km)*

State	*Loss*			*Gain*			*Net*
	Shifting cultivation	*Other reasons*	*Total*	*Net reg. in shiftin cultivation*	*Other reasons*	*Total*	*change*
Arunachal Pradesh	75		75	56		56	-19
Assam	257	159	416	163	16	179	-237
Manipur	603	-	603	463	-	463	-140
Meghalaya	75	2	77	20	-	20	-57
Mizoram	292	-	292	491	-	491	+199
Nagaland	573	-	573	503	-	503	-70
Tripura		3	3	4	7	11	+8
Total	1,875	164	2,039	1,700	23	1,723	-316

Source : State of Forest Report, 1997

The loss of forest cover in the region has reduced to 316 sq.km. in the 1997 assessment as compared to the 1995 assessment. *Jhum* or shifting cultivation is the single largest factor for the loss of forest cover in the region as shown in the Table 17.7. Destructive cutting of fuelwood has damaged the productive potential of open forest and scrubland. Unplanned and non-scientific commercial logging has also been destructive of the potential for regeneration of commercially viable species.

Judicial Intervention

There should be an effective legislation to ensure rational forest management so that exploitation forest is both economically viable and ecologically balanced. At present the existing legal measures for the forestry sector focus more on the productive aspect than environmental consequences. In practice the legal provisions for safeguarding forest resources and their scientific exploitation are considered inadequate. The Supreme Court of India had therefore to intervene in the matter and lay down guidelines for management of the forest with special reference to the north east region.

A writ petition (WP No. 171/1996) was filed in the Supreme Court by the Environment Awareness Forum Vs. State of J&K and others stating that the Khair trees in the state of J&K and in most other states were on the verge of extinction due to over exploitation, random felling and unscrupulous use by Khair wood based industries. Thereafter in May, 1996 the Supreme Court permitted the petitioner to amend the writ petition to cover felling of trees not only in J&K but in the entire country. During the course of hearings, the Supreme Court showed concern regarding excessive deforestation in the north east region of the country in particular, as revealed in the State of Forest Report, 1995 and overall low proportion of forest cover of the country (around 19.5 per cent against 33 per cent stipulated in the National Forest Policy). The Supreme Court issued notices to all the state/union territory governments to file affidavits.

Order dated 12.12.1996

After considering the affidavits of all the state governments, the central government and the views of the attorney general and the petitioner, the Supreme Court gave interim directions, as follows:

— The term "forest" covers all statutorily recognized forests, whether designated as reserved, protected or otherwise for the purpose of section 2 (I) of the Forest (Conservation) Act, 1980. The term "forest land" occurring in section 2 of the

Act of 1980 shall now include "forest" as understood in the dictionary sense, and also any area recorded as forest in the government record irrespective of the ownership. The provisions of the Forest (Conservation) Act, 1980 are to apply clearly to all categories of forests stated earlier.

— Total cessation of non-forest activities in forest areas including mining without prior approval of the central government.
— Felling of trees is to remain suspended in all the forests except in accordance with working plans approved by the central government.
— Complete ban on the movement of cut trees and timber from any of the seven north eastern states to any other state of the country either by rail, road and water except in case of certified timber required for defense or other government purpose.
— Complete ban on the felling of trees from the tropical wet evergreen forests of Tirap and Changlang districts in the state of Arunachal Pradesh.
— Court directed all the states and union territories to constitute expert committees to identify areas which are forests, as per its directions and also to assess sustainable capacity of forests of the state saw mills and timber based industry.

The Supreme Court passed specific orders for the state of Jammu & Kashmir, Himachal Pradesh and the hill regions of the state of Utter Pradesh and West Bengal based on the affidavits filed by the respective state governments.

Order dated 4/3/97

A High Power Committee (HPC) was constituted to oversee the implementation of the order or the Supreme Court. The HPC consisted of Shri T.V. Rajeswar, Shri. R.N. Kaul and a representative of the Ministry of Environment and Forests. The main functions of the HPC were as follows :

(i) overseeing preparation of inventory including timber and timber products lying in mill premises;
(ii) allowing use or sale of timber and timber products;
(iii) checking the illegality or otherwise of timber lying with wood based industries;

(iv) imposing suitable fines on timber based industrial units found to be involved in illegal felling (expo-facto approval given by the Supreme Court);
(v) regulating movement of timber products.

Order dated 8/5/97

All felled timber shall be inventorised and brought to the specified depots. After inventorisation, legitimate stock of timber lying within mill premises may be allowed to be converted by the HPC.

Order dated: 15/1/98

Realising the difficulties faced by the north eastern states, and various recommendations of the High Power Committee, the Supreme Court during January, 1998 sought the perception of Ministry of Environment and Forests, Govt. of India. The ministry taking into account all aspects stated that though proliferation of wood based industry may be the main cause of degradation of forests in this region, considering the extent of forests and dependence of local people on the forest resources, it is neither feasible nor desirable to completely ban either the timber trade or operation of the wood based industries. However, their numbers and capacities need to be regulated according the sustainable availability of forest produce.

The Supreme Court permitted utilisation of the felled timber and future working of forests in the following manner :

(i) Disposal of timber shall commence only after concerned principal chief conservator of forests irrevocably certified that inventorisation of all felled timber in the state has been completed.
(ii) All inventorised timber including seized timber lying in a forest should be immediately transported to specified forest depots.
(iii) The timber belonging to government and district councils shall be first offered for sale to government departments for their bonafide official use and rest shall be sold in public auction. Private timber owners whose stocks have been cleared by the High Power Committee shall have the option of selling the timber either in auction organised by the state forest departments/corporations or directly.

(iv) The timber including sawn timber can be transported only by the railways to other parts of the country. The MOEF has been empowered to relax the condition in specific cases.

(v) The state government shall formally notify industrial estates for locating the wood based industries in consultation with the Union Ministry of Environment and Forests.

(vi) The timber as per the inventories cleared by the HPC may be allowed to be converted/utilised if the unit is iocated within the notified industrial estate. However as the re-location in the proposed industrial estates may take some time, existing units with only legal stock may convert timber, a one-time exception, till such stock lasts subject to a maximum period as per the norms prescribed by the HPC or six months whichever is less.

(vii) State government shall ensure disposal of illegal timber before permitting the conversion/disposal of legal/authorized timber available with wood based industries.

(viii) There shall be a complete moratorium on the issue of new licences for the establishment of any new wood based industries for the next five years after which the situation shall be reviewed with the concurrence of the MOEF. All the wood based industries will be allowed to function in the industrial estates to be notified by the state government in consultation with the MOEF.

(ix) The future working of forests should be done only in accordance with the working plans approved by the Government of India and the working plans should be prepared within a period of two years. During the interregnum, the forest shall be worked according to the annual felling programme approved by the MOEF which shall be incorporated in the concerned working plan. The forest under the district, regional and village councils shall be worked in accordance with working schemes. The plantation schemes raised on private and community holdings shall be excluded from these requirements, but shall be regulated under respective state rules and regulations.

(x) An action plan shall be prepared for necessary patrolling and other necessary protective measures shall be undertaken in vulnerable areas.

(xi) The total sale proceeds from the sale of seized timber as well as timber products and penalties would be credited to the state revenues. Out of this, the state shall utilize one half of the amount for raising forest plantations by local tribal population and as assistance to the tribals. The remaining one half, after deduction of expenses, would go to the state coffers for other developmental activities.

From the details mentioned above, it would be seen that the Supreme Court has ordered that the working of a forest should be done in accordance with the working plan approved by the Government of India. The court reiterated that National Forest Policy for scientific management on sustained basis should be scrupulously followed. Considering the submissions of the ministry about the weak infrastructure in the north east and the fact that most of the forests in the north east were never brought under working plans, apex court granted a period of two years for preparation of working plans in the north east states. During the *interregnum,* forest shall be worked according to the annual felling programmes approved by the Ministry of Environment and Forests, which shall be incorporated in the concerned working plan. Thus, the Supreme Court prescribed mechanism according to which the forests can be harvested on a sustainable basis except the national parks and sanctuaries, which are to be maintained for ecological objectives. Such area is less than 5 per cent of the geographical area of the country.

The National Forest Policy, 1988 considers ecological security and meeting the needs of the local people as the primary function of the forests. The policy stipulates that the forest based industries should raise their own resources preferably by establishing direct linkage with individual farmers by providing inputs like credit and technical advice. Neither the forest policy nor the central government imposed any ban on felling of trees for timber provided the same are silviculturally available. The trees standing on the land farming part of national parks and sanctuaries, however, have to be dealt as per the provisions of the Wild Life (Protection) Act,

Keeping in view the orders of the Supreme Court, the Union Ministry of Environment and Forests have taken many initiatives for development of forestry in the north cast.

— Establishment of a north eastern cell in the MOEF.

— Establishment of the Arunachal Forest Protection Authority, and the Nagaland Forest Protection Authority under the Environment (Protection) Act, 1986 with the following functions:

(i) exercise the powers under section 5 of the Environment (Protection) Act, 1986 for issuing directions and for taking measures with respect to matters referred to in clause (i), (iii), (v), (vi),(x),(xiii) and (xiv) of sub-section (2) of section (3) of the said Act;

(ii) monitor and issue of necessary directions for the timely implementation of directives given by the Supreme Court in Civil Writ Petition No.202 of' 1995 from time to time, including and in respect of

(a) disposal of felled timber in the state;
(b) to determine the number of wood-based industries which can work in the state on a sustainable basis;
(c) approval of identified industrial estates;
(d) relocation and licensing of wood based industries;
(e) scientific management of forests;
(f) pricing of timber and fixation of royalty;
(g) action against those responsible for significant illegal felling; and
(h) any other matter pertaining to protection and management of forest and environment and working of wood based industries as may be referred to the authority by the Central Government.

— Establishment of industrial estates has been approved for the states of Arunachal Pradesh, Assam, Meghalaya, Manipur, Nagaland and Tripura. The state of Mizoram does not need any industrial estate.

— Working plans, schemes have been prepared for a number of forest areas in the region

— Proposed establishment of forest development authorities at the district level. The authority shall be under the chairmanship of the conservator of forests with the deputy conservator of forests as the member secretary. The schemes to be undertaken by the authority shall include afforestation, construction of roads, wells, small culverts and bridges, water harvesting schemes, alternate source of energy etc. Funds for these schemes may be made available by the central government.

— Proposed implementation of an integrated scheme for propagation, processing and marketing of bamboo in collaboration with the United Nations Development Programme.

Order dated: 13/1/2000

Inspite of the orders of the Supreme Court and establishment of Forest Protection Authorities in the states of Arunachal Pradesh and Nagaland, the illegal felling of timber in the north-eastern region continues. According to the affidavit filed by the Ministry of Environment and Forests on the 13th January 2000 before the Supreme Court of India illegal timber is still being transported from some of the north-eastern states. In fact 50 wagons containing timber were detected at the Tinsukia railway station in Assam in October, 1999. Most of the seized timber was of illegal origin and ownership of the said timber could not be established. The matter was brought to the notice of the Supreme Court which passed an interim order on the 1st November, 1999 that the fifty wagons shall remain seized till further orders. Another 28 wagons containing timber were seized at Rajpura in Punjab between the 29th December 1999 and 3rd January, 2000. Out of these 28 wagons, only 11 wagons have been claimed by the consignors and remaining have not been claimed so far. The timber contained in 11 wagons appeared legal and has been released. However, the remaining quantity of timber in 17 wagons is yet to be claimed and appeared to be illegal. Most of the quantity originated in the state of Arunachal Pradesh. On the 8th January, 2000 two rakes containing 66 wagons were detained at Nangloi railway station in Delhi. Nobody has claimed this timber till date.

The Ministry of Environment and Forests have suspected the involvement of some mills in the business of illegal timber and requested the High Power Committee to review the clearances given to the units which are suspected of being involved in transportation of illegal timber. Based on the above request, the High Power Committee has suspended the clearances of 60 units (57 in Arunachal Pradesh, 2 in Assam and 1 in Nagaland) on the 12th January, 2000.

The Supreme Court of India considered the affidavit filed by the Ministry of Environment and Forests on the 13th January, 2000 itself and passed the following orders.

(i) The Ministry of Environment and Forests may take various steps deemed necessary by it for appropriate investigation, storage, disposal etc., of the detained timber and also to carry out such actions in future for detention, seizure and investigation of timber which may include:

(a) seizure of timber during investigation and or confiscation of unclaimed timber or claimed timber for which complete details sought by the Ministry are not furnished within the stipulated time;

(b) directing state governments/railways or any authority/ consignee(s)/consignor(s) to furnish details/documents required for investigation;

(c) directing state governments/railways/consignee(s)/consignor(s) to keep custody of the tumber;

(d) disposal of seized or confiscated timber through auction/ sealed tender either directly or through state governments or any other agency;

(e) constitution of multi disciplinary team for carrying out investigations including from investigating agencies of the central and state governments; and

(f) issue of comprehensive guidelines for regulating movement of timber and timber products, standardization of transit passes; and

(g) reconciliation of movement of timber with its origin from inside the north-eastern state and outside these states.

The Supreme Court further directed that no other court will entertain any petition with regard to above mentioned cases and timber seized after the 13th January, 2000. As per the decision of the Supreme Court, the Ministry of Environment and Forests have created a Special Investigation Team (SIT) for detailed investigation for fixing responsibility and formulation of detailed guidelines on the 14th March, 2000. The SIT consists of Shri. S.C.Sharma, Addl. Inspector General of Forests, Ministry of Environment and Forests, Sh.G.K.Pillai, Joint Secretary (NE) in the Ministry of Home Affairs,

Sh.M.K.Jiwrajka, Head, North East Cell, Ministry of Environment and Forests and Sh.A.R.Chaddha, Deputy Inspector General of Forests in the Ministry of Environment and Forests.

The directions of the Supreme Court and the measures taken by the Union Ministry of Environment and Forests have helped in regeneration of forests in the north east India and also resulted in their conservation. Actual position will however be clear during the next survey of the forest resources.

Conclusion

As we approach the next millennium, the forestry sector faces increasingly difficult and new challenges. The socio-cultural benefits of forests and the social implications of the distribution of forest benefits are new receiving greater attention. Proper balance has to be achieved between the needs and the rights of the indigenous groups, forest dwelling and forest dependent population and range of demands for forest groups and services by a 'wide variety of interest groups. There are also additional considerations of population growth, changes in population distribution, economic pressures and efforts to alleviate poverty and ensure food security for all. This may lead to more intensive scrutiny of forests actual and potential contribution to development and of the relative benefits of retaining land in forests versus converting it to other land uses. Another factor of importance in this connection is the ability of forests to provide a range of environmental functions such as the conservation of biological diversity, mitigation of global climate change, protection against deserification protection of soil and water resources, protection of wild life etc. The greatest challenge in the forestry sector in the next millennium before us is to reconcile these considerations with safeguarding the rights of forest dwellers and indigenous people and to ensure wide spread participation in decision making in forest management.

REFERENCES

1 FAO. 1976. Development and forest resources in Asia and Far East Region - Trends and Perspectives 1961-1991. Food and Agriculture Organisation of the United Nations, 89 p.

2. FAO. 1990. The environmental impact of economic incentives for agricultural production: a comparative law study. Food and Agriculture Organisation of the United Nations, Legislative Study 38, 84 p.

3. FAO, 1994. Decline and die back of trees and forests— A global overview. Food and Agriculture Organisation of the United Nations, Forestry Paper 120.90 p
4. FAO, 1994. Forestry Education - New trends and prospects., Food and Agriculture Organisation of the United Nations, Forestry Paper 123, 305 p.
5. FAO, 1997, State of the Worlds Forests, 1997 Food and Agriculture Organisation of the United Nations, 200 p.
6. FSI, 1998. State of Forest Report, 1997. Forest Sur·ey of India (Ministry of Environment and Forests), 72 p.
7. S.0.827-(E) 1998 Order dated the 17th September,1998 issued by the Ministry of Environment and Forests regarding constitution of the Arunachal Pradesh Forest Protection Authority published in the Gazette of India: Extraordinary Part.II-Sec.3(ii) dated the 17th September 1998
8. Supreme Court Judgements

   (a) T.N.Godavarman Thirumalpad Vs. Union of India W.P. (Civil) No.202 of 1995 and no. 176 of 1996 Di..:ctions dated the 12th December 1996.
   (b) T.N.Godavarman Thirumalpad Vs. Union of India, W.P. (Civil) No.202 of 1995 Order dated tile 4th March 1997.
   (c) T.N.Godavarman Thirumalpad Vs. Union of India W.P. (Civil) No.202 of 1995, Environment Awareness Forum Vs. State of J&K and Other W.P. (Civil) No. 171 of 1996, Order dated the 15 January 1998.
   (d) T.N.Godavarmah Thirumalpad Vs. Union of India W.P. (Civil) No.202 of 1995, I.A. of 2000, order dated the 13th January, 2000.

18

CIVIL SOCIETY AND THE POLITICS OF PEACE-MAKING IN NORTH-EASTERN INDIA

— *Samir Kumar Das*

Twentieth Century has now reached a point where conflicts amongst various ethnic communities have decisively proven to be mutually detrimental to all those who are involved in them. This is what makes peace so much indispensable to us at the end of this century. This paper proposes to underline the importance of civii society in making and perpetuating peace in an otherwise turbulent region *viz.* the northeast. Such an endeavour is necessary at least on two counts: First, South Asia in general and northeastern India in particular have a very poor track record insofar as peace-making is concerned. While the communities cannot just afford to remain in what may be described as 'a war of all against all', persistent failures in making peace have turned out to be too expensive to put up with any longer. Secondly, all this happen, at a time when civil society in northeastern India - almost without any exception, is in direstraits. We have either not been able to organize it or syste matically eroded it wherever it has already been organized.

This paper is founded on two rather inter-related assumptions: One, the erosion of civil society has a direct bearing on the persistent failures of peace accords in the region. The paper in fact argues that any attempt at making peace by way of decimating the civil society is bound to end up in a fiasco. The point may be pushed a step further in order to drive home that restoration of civil society is absolutely essential for making peace accords successful and achieve their intended end. Two, we also assume that peace is critically contingent on a particular kind of accords specified later, rather on any and every accord. Peace for us, is inconceivable without the instrumentality of accords. Accordingly this paper is divided into three parts : In the first part, we analyze the relevance of civil society to the process of making peace. The second part discusses how the state has in some cases, abstained from constructing the civil society and in others, sought to scuttle and paralyze it. The third part focuses on some of the proposals of the insurgent as well as non-insurgent organizations mainly based in Assam for resuscitating the civil society institutions in the region.

Accords as the Instruments of Peace

We look upon accords as important instruments of peace. We think, this is a somewhat blank statement that has to be qualified by at least two riders : One, accords are not the only instruments of peace. The state may believe that military means may be much more useful than anything else at any given point of time, though of course, peace achieved thereby if at all, is unlikely to be of lasting and durable nature. Two, that all the ethnic peace accords have failed in the northeast - albeit, with same rare exceptions like, the Mizo Accord of 1987, doss not mean that accords are destined to fail in the region. It is the failure of a particular kind of accords and not accords *per se*. The way accords are signed and executed is just not expected to deliver the goods and facilitate the peace process.

In this connection, we may raise two closely linked questions. How do accords serve as important instruments of making peace ? And, in what sense, do they fail in becoming the instruments of peace? Accords in Classical Liberal Theory cantain three ingredients : First, the signing of accords is necessarily preceded by discords. Unless there are discords, there seems to be no necessity of signing the accords. Discords as the theory envisages them, arise out of conflicting interests of the actors involved in a given situation. Although the theory views the individuals as potential holders of interests conflicting with each other, the same view can be adapted to the collective sphere where groups or communities are seen to hold conflicting interests. Two, any transference of the Classical Liberal Theory to the collective sphere of groups may be extremely difficult for although it envisages the presence of a multiplicity of communities conflicting among themselves, the theory assumes that they operate within an abiding and commonly agreed upon framework of (nationalist) consensus. In other words, community conflicts are necessarily tempered by a consensus that the conflicts - however serious they are (i) do not threaten to encroach on the right of others to hold and pursue their respective interests and (ii) that the pursuit of their own interests requires the mediation of others. John Stuart Mill for instance, argues that the representation of their interests is possible only within a uniform social and normative order of the nation. As we will see later, the role of civil society in arbitrating and reconciling the conflicting interests can hardly be exaggerated. Three, accords are held as 'rational' instruments on the ground that the signatories be way of entering into them feel that their interests can be better secured and served by way of entering into them rather than by being otherwise mired in an endless series of discords. Or it may be that by way entering into accords, they feel that they can keep the loose or damages to a bare minimum or get them equally distributed. In a trade-off between accords and discords, communities left to themselves, are likely to choose accords. Liberal Theory does not recognize those as accords in which one party is coerced or pushed to a corner where it is left with no option but to sign them without making any rational calculations.

State, Civil Society and Peace

We regard civil society as the cornerstone of peace : whether in South Asia or in India, primarily on three counts : First, it is a sphere that is expected to enjoy a certain degree of independence from the state in the sense that it is not consumed and obliterated by the state. If its independence is jeopardized in any manner, accords will be reduced to what Ranabir Samaddar calls, 'those documents and are unlikely to be rational instruments: Civil society is a concept that is used to refer to this independent sphere- considered by the Liberal Political Theory to be necessary for making accords successful. Secondly, civil society sets forth the terms of a broad, overarching consensus within which conflicting interests are held and pursued by diverse ethnic communities. While a community can hardly be seen as a monolith speaking in one voice, civil society consisting of a multiplicity of ethnic communities is unlikely to be permeated by any overwhelming interest. The society allows for circulation and fermentation of interests not only amongst communities but also within communities. Debates and discussions therefore are central to the constitution of any civil society. This is what Craig Calhoun calls, 'discourse over social arrangements'.[2] Civil society being in a state of fermentation and flux, cannot allow any of its actors to have the final say in this respect. For what I consider to be my interest or the interest of my community gets modified or even transformed by the debates and discussions that characterize it:' Thirdly, the actors are not only independent but also rational. To say that one is independent is not the same as saying that one is rational too. Civil society is usually interpreted as the nursery that imparts to the potential actors the requisite awareness training. One is trained not only to discern what one's enlightened self-interest is, but also how to pursue and attain its by way of exposing oneself to the influence of others. Civil society enables the independent actors to temper their interests with rationality.

While civil society is the cornerstone of peace-making in the northeast, the entire region has been a standing witness (to a process) of systematic erosion of civil society institutions right from the colonial times. Nowhere is this more glaringly illustrated than in the state's endeavours at making peace and getting the accords signed while simultaneously paralyzing the civil society. A study of the Shillong Accord (1975)—widely regarded as the precursor to many of the subsequent peace accords in the northeast, may be instructive in this context. It was signed by persons acting on behalf of the Government of India and 'representatives of the underground organizations' respectively : First, the signature of the accord was preceded by a spate of fierce military operations undertaken by the state till the underground forces were considerably weakened with the effect that they were left with no option but to capitulate and surrender. This is borne out by a comment of Y. D. Gundevia, then the Foreign Secretary of Government of India that the Government should negotiate from a position of strength and should wait till they get an upperhand to dictate terms. Secondly, Government's attempts at driving a wedge within rebel ranks by way of forcing splits in their organization are much too well-known now. The traces of fissures within the ranks of Naga National Council (NNC) were reportedly visible first in 1956 when Mr. Sakhrie was mysteriously murdered. The conflict between Naga National Council/Federal Government of Nagaland and Council of Naga People/Revolutionary Government of Nagaland is a case in point. Thirdly, the state also created a space in the over ground in a bid to wean away the moderates from the rebels and allow them to enjoy a share in the country's power structure and most importantly, offer to them wide-ranging political concessions including autonomy and statehood within the Indian Union short of sovereignty. The birth of the state of Nagaland may be interpreted along these lines. It was intended to be a safety valve much in the same manner as Indian National Congress was, expected by the colonial authorities to be one staving off the anti-colonial forces in pre-Independence India. As Subir Bhaumik writes : "The creation of Nagaland as separate state was intended to take the sting out of the rebellion—it was to put in place a new moderate leadership which would accept the Indian Constitution and its control over Nagaland for substantial concessions in terms of autonomy.[4] Fourthly, the state nevertheless engages itself in dialogues, debates and discussions but does so in a way that aims at weakening if not decimating the civil society. It usually takes inordinately long time while engaging them in talks and thereby wear them down. The whole idea is to continue the talks indefinitely without delivering the goods and the longer they talk without delivering anything, the more their credibility gets undermined in the eyes of their own people. Moreover, while conducting talks, the state is guided by what may be called, a piecemeal approach to what is otherwise considered to be an integral and comprehensive problem. President's rule was clamped on Nagaland on 25 March 1975. This according to S. C. Dev, then a Deputy Commissioner posted in Nagaland, "created for the bureaucrats opportunities to show their determination and commitment to principles, ideals and values." They adopted what Bhaumik calls, a tribe-by-tribe approach. The idea was to depend on the intermediaries like, the gaonburas, village elders, the dobhashis, the leaders of the public and family members of the hostiles and to ask them to get in touch with the underground -- failing which the state would unleash repression against them, and finally, to get them to surrender by way of accepting demands which are essentially of local nature, like, construction of metalled roads, bridges, rural employment etc.[5] The exercise is intended to fragment and thereby destroy what once was a comprehensive agenda of self-determination.

Proposals for Reorganizing the Civil Society

In this section, we confine ourselves to an analysis of some of the proposals of a few leading non-state organizations operating in the region for reorganizing the civil societies in the northeast in the face of such assaults on them outlined above. We will do well to keep two things in mind in this connection : One, we concentrate mainly if not exclusively on the proposals as we are able to decipher them from some of the writings of these organizations published in the local newspapers. Our focus is on the proposals rather than on their attempts at implementing them. The gap between them may be too yawning to be effectively bridged. Secondly, for our convenience, we propose to restrict ourselves to the experience of early 1990s - a time when reorganization of the civil society was made difficult by a series of army operations especially in Assam. The United Liberation Front of Asam's (ULFA) initial reluctance to frontally take on the army even before people are prepared for it, is much too well-known to be recounted here. But as soon as the operations began, the organization was obviously drawn into the ear'.

We may identify three major planks of reorganizing the civil society in their proposals : First, we all know that debates, discussions and dialogues are central to the constitution of the civil society. Their openness to these debates, discussions and dialogues during the period under review may be explained by a combination of factors : First, these enable the communities to shuffle and articulate their views and interests whether generally or on any particular matter. Given that a community seldom resembles a monolith having one view or only one single interest, these communicative processes go a long way in forming well thought out opinions and sharpening its interests. The role of some of the Naga Organizations in facilitating these processes cannot be underestimated in any manner.[6] The will of the community is neither that of the organizations claiming to represent it, nor even that of the majority, It is what gets articulated through such processes as debates, discussions and dialogues. Secondly and as a corollary to it, these processes give the organizations an opportunity of establishing contact with the people, of assessing their mind and being enriched by their opinions. One may recall that there were many weeklies or bi-weeklies in Assam which contained columns in which some of the top leaders would get an opportunity of responding to the questions of the ordinary people. Thirdly, these also help them in strengthening their organizations and in acquiring necessary theoretical knowledge. Exposure to public criticisms of constructive nature is a prerequisite for theoretical knowledge.

Secondly, these organizations sometimes have encouraged development activities in areas where the effects of state-initiated development have not really trickled down. Such activities are expected to have two-fold impact of making the community self-reliant on the one hand and filling up the void created by the absence of state-sponsored development processes on the other. The role of the Jatiya Unnayan Bahini in matters of development may be cited as a case in point. We have already seen that autonomy and self-reliance on the kingpin of reorganizing the civil society.

Finally, the reorganization of the civil society also presupposes the discovery of an appropriate principles on the basis of which civil society may be rebuilt. If the principle is not appropriate, then it might hasten its decay. ULFA's gradual realization that reorganization of civil society in Assam is critically dependent on its ability to retain her composite and multi-ethnic character seems to have made it distance from the early thesis of 'de-nationalization'. The plea for de-nationalization is double- edged : More negatively, it refers to a process whereby ethnic or tribal identities cease to lose their significance in social and economic spheres of one's life. ULFA is supposedly the microcosm of a de-nationalized body in which members are drawn across ethnic, tribal or Varna groups and mixed together without any trace of these prejudices and identities. It seeks to create an Assam after its own image. In more positive terms, economic identity prevails over all other forms of identity so much so that people may organize themselves on the basis of their economic identity and polarization of classes reaches a point where revolution becomes the vehicle of a new social order. It envisages a situation which the schism between *swadeshi* and *bideshi* is replaced by a contradiction between the exploiters and the exploited. The argument is that an emphasis on economic identity cleanses economic contradictions of what a commentator calls, 'social distortions and paves the way to social revolutions.[8] We will not go into the details of the criticisms that the thesis was subjected to at that time. What is important to note is that the thesis gradually gave way to a new principle of reorganizing the civil society in Assam. It obviously speaks of the dynamic nature of the organizational thinking. On the/one hand, it is evident that the organization is blatantly opposed to the hegemonism of the Assamase-speaking Varna-Hindus or for that matter, hegomonism of any other community. Most of the social movements in Assam since Independence including the Assam movement of 1979-1985 were the ones in which hegemonism of one community or another over others within an otherwise composite state like, Assam, has been the defining principle. Assamese hegemonism' (*Asomiya Pradhanyabad*) has according to the organization, done greater harm to the society of Assam than anything else. On the other hand, while a common fight against the enemy may be politically expedient, each of these communities according to it, is entitled to the right to self-determination after the fight is over : "We recognize each community's (*janagoshthi*) right to self determination based on equal rights and equal dignity in Independent Assam". The civil society in other words, cannot be grounded in any hegemonic principle. Democracy and civil society are inseparable from each other.

Peace in short, in today's northeastern India is critically dependent on reestablishing and reorganizing the civil society. Haw long are we going to take to appreciate their connection and lose at our expense, the battle for peace ?

NOTES

1. Ranabir Samaddar, 'Those Accords : A Bunch of Documents', SAFHR Paper Series - 4 (Kathmandu : South Asian Forum of Human Rights, 1999).
2. Craig Calhoun, 'Nationalism and Civil Society' in Craig Calhoun (ed.), *Social Theory and the Politics of identity* (Cambridge : Blackwell, 1994), p. 327.
3. Jurgen Habermas, *The Theory of Communicative Action*. Vol. I: Reason and Rationalization of Society London : Heinemann, 1984, p. 10.
4. Subir Bhaumik, 'The Accord That Never Was : A Critique of the 1975 Shillong Accord' (1999), *mimeo*.
5. S.C. Dev, Nagaland : *The Untold Story* (Calcutta : Gauri Dev, 1988), pp.98-114.
6. For example, Naga Mothers' Association, Naga People's Movement For Human Rights, Naga Hoho etc.
7. *Janamat* in *Budhbar* provides an excellent example.
6. Rajiv Sharma, 'Need for a Fundamental Approach to Insurgency' in Bolin Hezarika (ed.). *Socio-Economic and Political Change* : Some Issues : Jorhat : J.B. College, 1998.

19

NORTH-EAST INDIA IN THE 21ST CENTURY WITH SPECIAL REFERENCE TO NAGALAND

— *Sukhendu Mazumder*

North-East India, a land of abundant resources at her disposal, has not yet been able to forge ahead in the fields of economic development and poverty alleviation-cum-employment generation is certainly a matter of grave concern. It leads one to believe that it's her lack of political will in one side and of lack determinatoin coupled with complacency on the part of the people. It is also quite bewildering that there has been government units, semi-government institutions and NGOs which often come to the rescue and have been putting ceaseless efforts to set things right and still the outputs have not been to the desired extent. Hence, it can be viewed as a lack of zeal and fury on the part of the people in general. The same can be nicely traced in the remarks of Jean Dreze and Amartya Sen in their book entitled, "India: Economic Development and Social Opportunity". They write, "Successive governments in India have had reasons enough to rely on the unending patience of the neglected and deprived millions in India, who have not risen in fury about illiteracy, hunger, illness or economic insecurity". Toleration is a virtue and we Indians as a whole have certainly championed the issue and the world believe that it is worth following. However the noted economist J.M. Keynes has observed that in the long run we are all dead.

If we look into the prevailing state of affairs of the sub-continent as a whole, we of course do not notice anything promising either. Till date, more than one third of the population has been under the grip of poverty. It is also found that about 145 millions of Indians have no access to health services, more than 178 million people fail to get safe water and the same has been increasing to an alarming proportion also. Though there has been mushroom growth of educational institutions and such related facilities available in every nook and corner of the country and yet 250 millions adults are illiterate along with a huge number of drop-outs at the primary stage. In spite of the fact that we have made spectacular progress in the field of medical sciences, but in the record of infant mortality rate we are among the highest in the world.

However, in the rate of literacy growth the NE States do not fare that had and also in the employment rate the figure is higher than the All India figure. The comparative figure in the two fields are as follows (*Indian Economic Survey*: 97-98) :

Literacy Rate (1991)		*Employment Rate (1991)*
Punjab	58.51	30.87
Andhra Pradesh	44.49	45.05
Kerala	89.79	31.43
West Bengal	57.70	32.15
NE states		
Assam	52.90	46.24
Manipur	59.89	42.18
Mizoram	82.47	48.19
Nagaland	61.05	42.67
Meghalalya	49.10	46.24
Arunachal Pd.	41.59	46.24
Tripura	60.40	31.14
All-India	52.99	37.42

But the figures here are not to be misunderstood as the correct reflections of the position in the real sense of the term. The entire North-East does not have any major industry to speak of baring the Tea-Estates and Oil-Refineries. This has virtually crippled the states to generate any viable source of employment. The Chief Minister of Nagaland rightly observes at the Chief Ministers' Conference held at New Delhi on 14 Sept. 1998. On the issues in the agenda for today's meeting, Nagaland and the North-Eastern States barring Assam, may not have much comment since ours, are all nascent and under-developed economies. Our development in the last fifty years has been focused mainly on the provision of infrastructure and basic amenities which were non-existent when the States were created. We therefore do not have any industrial base worth the name and therefore, our production of goods is and the scope for taxation is very limited.

Since the inception of the state, we have been trying our best to diversify and enhance our tax revenues. However, till today, with total prohibition being enforced in the State in pursuance of the directive principles enshrined in our constitution, our tax revenues are mostly from sales tax and motor vehicles tax, along with a nominal land revenue and house tax. "Hence, to accelerate the pace of economic development viz-a-vis employment, emphasis should be laid on the small-scale agricultural sector and labour intensive industrialization of the entire zone.

In this regard, I would like to remind the intelligentsia that we must take proper care of the tribal population who contribute more than 8 per cent of the populace of the country. The indigenous people of the land known as the Scheduled Tribes live with a unique harmony with the nature and they grossly depend upon the forests and agriculture. So a right balance is required between the process of modernization and the traditional systems. It is understood that the Govt. of India has been putting efforts for their economic development in a sustainable manner. In the new Millennium, we have to frame things in such a judicious manner that the nation's ecological heritage is given due importance.

Sl. No.	*Name of the States*	*% of Women employment*
1.	Arunachal Pradesh	10.91
2.	Assam	5.94
3.	Manipur	16.03
4.	Meghalaya	16.03
5.	Mizoram	13.05
6.	Nagaland	9.59
7.	Tripura	8.50
8.	All-India	16.59

Hence, such plans and programmes if not implemented to the fullest extent, we can't expect a change in the plight of the people.

In Nagaland, there has been lots of factors playing to determine the situation. Here the entreprenurial activities often receive threats due to unreasonable extortion's which the Govt. machineries fail to combat resulting into flight of the capital from the State. Obviously, this keeps the people poor and there is virtually no way out also. There are many NGOs in the State trying to safeguard the interest of the Naga people always, but they have not been that successful either in their mission. Prominent among them is the Naga Peoples Movement for Human Rights (NPMHR). The NPMHR has always been active in the protection and safeguarding the human rights violation in the area here. The NPMHR has always remained vocal against the wrath of the forces which often cause immense damage to the people who are innocent men and women, and children. It is of course a good sign that the underground forces are gradually being brought to the peace talks and there is hope emerging that Nagaland may ultimately see the better day ahead.

Now let us examine the resource position of the State. However there has been political unrest in most of the NE states for autonomy (specific mention may be made of Nagaland and Manipur) and yet the economic issues etc. have not been targeted to the extent they deserve. Now the time has come and we are in the twenty first century where the world has entered into the computerized life style. We Must now look into the prospect and chalk out a clear plan to give this zone a healthy look. "The remote and backward regions of the North-East are being totally left out of the processes of advances in information technology and thereby we are nowhere on the information highway nor part of the information revolution taking place all round the world. We are also being unable to get advantages of the special concessions being given by the Government of India to this sector. I request the Finance Commission to provide a special chapter on the need to upgrade the computer and electronic technology in the North-East by provision of adequate means for upgradation of systems both in the Public and private sectors. I feel that this will help in our attempt to rise to the levels of modernization and development with other parts of the country." (excerpts from the speech of the Chief Minister of Nagaland at the meeting with the Eleventh Finance Commission at Kohima on the 17th May, 1999).

In Nagaland, we have 'reserved forest' area at the tune of 8593(ha), purchased forests 29247(ha, protected forests 50756 (ha), wildlife sanctuary 22237(ha), and in village forests there are: i) Virgin forests at the tune of 477827(ha) and ii) Degraded forests of 264280 (ha). Thus it is found, that out of total land of 16,57,583 hectares, the forests occupy an area of approximately 8,62,930 hectares. Comparing this, it is clear that Nagaland has failed to tap and utilize the resources properly towards the economic development of this hilly state.

In the fields of irrigation, it is found that the irrigation potential created is at the tune of 65.527 ha against the total of estimated irrigation Potential available at 179000 ha i.e. 36.60 per cent. From the 7th plan onwards the involvement of the village community towards rural development has become a reality. In 1989 Jawahar Rojgar Yojana and in 1993 employment assurance scheme were introduced in the state. To remove poverty, IRDP. TRYSEM and DWCRA etc. have been undertaken also. Along with these, to help the people in their efforts to economic development, Basic Minimum Services (BMS) has come into force since 1996-97. In addition,, there are many more schemes undertaken for this state; some of which are: Ganga Kalyan Yojana, Million Wells Scheme etc. etc.

In the fields of agriculture, there are lots of schemes including 16 centrally sponsored schemes and NEC schemes. But till date the state is miles away from self-sufficiency. During the 8th plan period 28 water-shed projects were completed under the National watershed Development Project in Rainfed Areas and in the 9th Plan the number has increased to 81 and the works will continue for these three years more (i.e. 2001-2002). There is a very good possibility of Tea Plantation also in the state and farmers are being assisted to develop interest in it. So far 3.5 lakes tea colons have been raised and distributed during the 97-98 period.

In Nagaland, the Village Councils are working at the bottom rung and the District Planning Board at the district level. But the District Rural Development Agencies were set up only in July 1995. But the plus point is that the Deputy Commissioners are to function as Chairman of both the bodies leading to an effective liaison between the DPB and the DRDA. However the functioning of the DPB and the DRDA requires speed and accuracy based on experience of the sister states.

Thus it can be hoped that the New Millennium may really begin a new phase in the life of the entire North-East provided things are done in right earnest.

REFERENCES

1. Udayon Misra-North-East India (Quest for Identity. (88, Omsons)
2. Jean Dreze & Amartya Sen-Economic Development and Social Opportunity - (OUP, Delhi, 95).

3. The Naga Imbroglio, Charles Chasie (Standard Printers & Pub. Kohima, (99).
4. N. Sanajaoba, Human Rights Ed., 'Omsons'
5. G.K.Ghosh-Tribals and Their Culture.
6. R. Gopalakrishnan - Ideology, Autonomy and Integration in the North East India.
7. The Historical Memoranda of the Nagas and their Neighbours - Mar Atsongchanger.
8. Pravas Rn. Bhattacharjee - *Economic Transition in Tripura* - (Vikash Pub., New Delhi, 1993).
9. Paul Manansale -The Naga Race - (Pub. Firma KLM Pvt.Ltd. Calcutta, (94).
10. Speeches of S.C. Jamir, Chief Minister, Nagaland, Pub. by Dir.of Info. & Public Relations, Govt.of Nagaland, '97, 98 &99.
11. Trifed- Ministry of Social Justice & Employment, Govt.of India.
12. Materials on Rural Dev. Prog. in Nagaland. (Govt. of Nagaland).
13. *Journals of NEICSSR* Apr. '99, Oct.'99, Dec. '99.
14. *Manipur Today* - 1999, Dir. of Info. & Pub. Relations, Manipur.

20

HUMAN RIGHTS WITH SPECIAL REFERENCE TO NAGALIM

— *Xavier P. Mao*

The talk of human rights became very pronounced and much discussed as topic after the French Revolution in 1789-1799 and the American War of Independence in 1775-83. It assumed greater significance in the context of contemporary society after the Universal Declaration of Human Rights by the United Nations, on 10th December 1948. Further it was widely appreciated, deliberated and discussed after the formation of Amnesty International in 1960 and the Policy Statement on Human Rights by the National Council of the Churches of Christ in the United States of America, on December 6, 1963.

What is Human Right?

The basic philosophical question about Human Right is what is the basic concept of Human Right. Who is the authority of this Right? Who has confessed this Right on human being? The Human Rights are not legal rights conferred by politico-legal authority. These are rights by virtue of which human beings are human beings. In other words, they refer to the inherent dignity and of the equal and inalienable rights of all members of the human family as the foundation of freedom, justice and peace in the world. Take away these rights, human beings will be dehumanized. In a nutshell, Human Rights are a species of Natural Rights. These rights cut across the boundaries or limits of race, language, culture, sex, religion etc. As a matter of fact there is a religiomoral assumption behind this understanding. According to Christianity, we human beings are all children of God. According to Upanishadic-Vedantic tradition of Indian Philosophy, we human beings are essentially the manifestation of one eternal Brahman. Further in affirmation of the religious view of humans,

German philosopher I. Kant asserted that all human beings should be treated as ends and not merely as means. Kant offers us valuable insights as to how we ought to deal with other humans. It seems to me that there is considerable moral progress to the extent that we recognize that other people do have interests. The various excuses often invented for excluding others from moral consideration, partially or wholly, do not withstand rational scrutiny in the ultimate analysis. In other words, we can be said to have made definite moral progress only if we progressively generalize our rules like the Kantian categorical imperatives. This is moral progress because we are transcending the particular body of interests presumably those interests most meaningful to the rule-framers. Thus moral progress consists in the respect for the interests of others, although the interests are varied and many. To say that our moral and legal codes are better and refined is to say that human rights in general are also better, and more refined. Since human rights entail morality. As often when we treat other people wrongly, this is very often because we do not care about acting morally or because we do not recognize the others affected as having moral status or because we are ignorant of the effects of our actions on them. The moment we recognize that other people do have moral standing, then we are well on the way toward treating them decently.

The moral insight Kant has given is that human morality is to recognize that other people really do count morally and count fully. Although most people hold at least in lip service that all human beings count morally, but this has not always been the case, nor is it now always the case that others are accorded the moral significance they ought to be accorded. It seems to me that there is clear connection between the respect for human rights and the moral progress. If we act with only minimal good will toward others, we will avoid by far the worst crimes of human rights violation. The worst excesses of human rights violations do not stem from imperfect ethical systems. They spring from people not being concerned to act morally toward others and from people not having the least good will toward others. For example, the blacks in USA and in Europe and the low castes in India are not treated with moral consideration. Further the Drug Companies dump dangerous drugs on third world countries because they care more with their profits and not because they are ignorant of the consequences. Right from international terrorism, to genocide, to the episodes of petty nastiness or simple in consideration of day to day life, we frequently fail morally because of the lack of good will toward some particular group or all others. If we were to accord moral significance to others and acted toward them with good will, and tried intelligently, to arrange our social institutions accordingly, we would still not achieve perfection as we are imperfect beings, but we would surely do very well.

The so called human rights enshrined in various constitution of the world are derived from and justified by these basic rights. Next I wish to argue out the human rights enunciated by the United Nations. And in general how human rights are seriously violated through out the world and in particular in Nagaland or Nagalim. It is stated in the; preamble to the Universal Declaration of Human Rights that disregard and contempt for human rights have resulted in barbarous acts have outraged the conscience of humankind; and the coming of a world in which human beings shall enjoy freedom of speech and belief and freedom from fear and want has been proclaimed as the highest aspiration of the common people.

Other important constituents of the Preamble to the UN Universal Declaration of Human Rights is reproduced below:

> Whereas it is essential, if men is not to be compelled to have recourse, as a last resort, to rebellion against tyranny and oppression, that human rights should be protected by the rule of law, whereas it is essential to promote the development of friendly relations between nations,
> whereas the peoples of the United Nations have in the Charter reaffirmed their faith in fundamental human rights, in the dignity and worth of the human person and in the equal rights of men and women and have determined to promote social progress and better standards of life in larger freedom, whereas member States have pledged themselves to achieve, in cooperation with the United Nations, the promotion of universal respect for an observance of human rights and fundamental freedoms, whereas a common understanding of these rights and freedoms is of the greatest importance for the full realization of this pledge,
> Now, therefore, the General Assembly proclaims this universal declaration of human rights as a common standard

> of achievement for all peoples and all nations, to the end that every individual and every organ of society, keeping this declaration constantly in mind shall strive by teaching and education to promote respect for these rights and freedoms and by progressive majors, national and international, to secure their universal and effective recognition and observance, both among the peoples of Member States themselves and among the peoples of territories under their jurisdiction.

From the various reports and documents that I came across from various parts of the world as a member of the Amnesty International for three consecutive years, I find that the frequent violations of human rights are with some specific articles of the UN Universal Declaration of Human Rights. The specific articles are given below:

Article 3: Everyone has the right to life, liberty and security of person.

Article 5: No one shall be subjected to torture or to cruel inhuman or degrading treatment or punishment.

Article 6: Everyone has the right to recognition every where as a person before the law.

Article 9: No one shall be subjected to arbitrary arrest, detention or exile.

Article 18: Everyone has the right to freedom to thought, conscience and religion; this right includes freedom to change his religion or belief, and freedom, either alone or in community with others and in public or private, to manifest his religion or belief in teaching, practice, worship and observance.

There is the apparent paradox involved in the concept of human right. The concept of human right in its fullest logical extension appears to lead to solipsistic individualism and negation of community or society or collective life. My conclusion will be, only when human rights are safe guarded, then only community or social or political life will be possible and stable. Human right is the basis of legal, political, social, religious and economic rights. In other words, the basis of community, polity and society is human right. Thus human rights are not conferred by any legal authority. In pursuance of human rights, non-violence can be the effective method. Violence as a token and not as a creed can be used in certain cases,

When I reflect over the human rights situation in Nagalim what immediately strike my mind is the gross violation of human rights by the armed forces of the state machinery. And to a lesser extent by the so called "freedom fighters", the obvious examples are the shooting of innocent people, disappearance cases, rape, cruel and brutal method of interrogation of both civilians and 'underground', forced labour, arbitrary detention, desecration of churches, destruction of villages, crops and other properties, by the armed forces. The subhuman treatment meted out to both civilians and 'underground' is appalling. Such beastly treatment paved the way for a society of psychologically abnormal people, and violent and warped personality, of a large number of people in a society. The severe restriction of foreigners to visit the North-East, especially in Nagaland and Manipur itself is a serious violation of human rights, i.e., freedom of movement and association. This also reflects the inefficiency and the non-functioning of the various intelligence agencies of the state. Thereby, innocent foreign tourists, scholars, foreign spouses of native North-East, especially from Nagaland and Manipur, are being subjected to absurd and unnecessary periodic report to the nearest police station, even if they are allowed to visit the region. Even the various militant outfits operating in Nagalim are fast losing public support because of their inconsistent stand assumed by them from time to time. The Naga public is constrained to express their views and opinion in this regard. The indiscriminate search of the armed forces to the innocent passengers is also a violation of one's privacy and which amounts to undue harassment to the public. The so called 'freedom fighters' treatment to the informers are also equally cruel. The number of people killed within the factional clashes are very high. The so called tax collection and extortion by the various factions from the public itself is a kind of harassment to the public. All these are instances, to my mind, serious violations of human rights apart from many other violations.

It seems to me that the effective solution to the Naga insurgency problem is the recognition of the genuine and deep problem and to initiate a serious dialogue and negotiation. Treating the other as object and not as subject from either side is bound to the inevitable continuous violations of human rights. Therefore, greater militarization on the part of the state machinery and intensified activities on the side the insurgents are only paving the way to the greater violations of human rights.

REFERENCES

1. *Encyclopedia of Philosophy*, Vol. 7 & 8
2. The range of Ethics by Harold H. Titus/Morris T. Keeton, 1972, Litton Educational Publishing Inc.
3. A Morally Deep World by Lawrence E. Jhonson, Cambridge University Press, 1991.

21

REGIONAL DISPARITIES IN ECONOMIC DEVELOPMENT

An Inter-District Empirical Study of Arunachal Pradesh

— *Amitava Mitra*

Balanced economic development of different regions or States of a large country like India is of great importance, not only from the economic point of view but also for political stability. India launched the process of development through planing as a development strategy since 1951. The principal objectives of economic planing are economic growth, self-reliance and social justice which includes better interpersonal and inter-regional income distribution and also balanced regional development. There is no doubt that during the period of more than four decades, of economic planning. the country has been able to make remarkable and significant achievements in various sectors of the economy. However, most of these achievements are more or less at a macro level and when we come down to the micro level, or to be specific, to the regional or state levels or to levels further down like district and block levels, much is yet to be done to achieve a meaningful economic development based on social justice and balanced regional development. Although the problems of development of backward areas in general, and hill areas an particular, received increasingly higher attention from the central planners since the Fourth Five Year Plan but the emphasis on the preparation of district and block level plans have been laid only during the late seventies and early eighties. In order to make future plans for the districts, blocks, and even for the circle levels, it has become essential to study the socio-economic structure at the district, block or circle levels. The identification of backward districts is the first step towards the successful implementation of the plans to attain balanced regional development of the country as well as the State. Hence, in this paper an attempt has been made to study the socioeconomic structure of the various districts of the State and to identify the relatively backward districts of Arunachal Pradesh.

In this study we would like to focus on the levels of development at the district level i.e. more specifically to identify rural families identify the relatively backward districts. There are four stages involved in this task and at each stage one cannot help making & series of judgments. These stages are(a) Selection of indicators (b) decision about unit area for which data is to be collected (c) compositing the indicators and (d) classification on the basis of homogeneity.

Selection of Indicators

Development and backwardness are the two sides of the same coin. As development/backwardness is a multidimensional phenomenon and the levels of development/backwardness are not directly measurable. we have to select suitable indicators. A development indicator, should represent some aspect of development such as industrialization, health, education participation etc. because development involves changes in structure, capacity and output. It may be a direct measure of economic or social variable or more often an indirect measure of some non- measurable phenomena i.e. standard of living. Hence, indicators are generally in the form of derived summary figures such as averages, rates, ratios and other- synthetic functions.

The Committee on Dispersal of Industries (1960) suggested three indicators of backwardness, names (a) poverty (b) high population density in relation to available productive resources and (c) lack of communication. As a follow up of the Committee's Report, a working group was appointed by the Planning Commission in the context of the Fourth Five Year-Plan to identify backward areas of states which were classified into five categories namely (a) desert areas (b) chronologically drought-prone areas (c) areas with high concentration of tribal population (d) hilly areas including border areas and (e) areas with high density of population and low level of income. Most States of North-Eastern Region of India including Arunachal Pradesh fall in the (c) and (d) categories. The working group recommended around fifteen indicators of backwardness/development which were accepted by the Planning Commission. However, in Arunachal Pradesh. some of the indicators like the value of agricultural produce, are not available at the district level nor at the state level. Further, some of the indicators like the number of workers employed in the manufacturing sector, number of commercial vehicles registered in a district etc. do not have relevance in the context of Arunachal Pradesh. Similarly the six indicators suggested by the Panda Working Group for identifying backward districts within the State are not found to be applicable even today in the case of districts of Arunachal Pradesh, due to non-availability of data and also non-relevance of some of the indicators like per capita income from industry and mining, number of workers in registered factories, railway mileage etc. Hence the analysis of these reports suggest that the indicators should be carefully selected and they should be based on availability and comparability of data as well as relevance of data in the context of a particular region or a State.

On the basis of the availability and comparability of data at the district level, twelve indicator were carefully selected and broadly classified into three main heads like (a) economic indicators (b) social indicators and (c) infrastructural indicators.

Economic Indicators

As agriculture is the most important occupation and the majority of the people are engaged in agricultural occupation in Arunachal Pradesh the level of economic development can beat be reflected in agricultural efficiency. Hence, out of the five economic indicators selected there are four indicators which are related to the agricultural sector. The last indicator is also related to rural areas since around 87.20 per cent of population in Arunachal Pradesh live in the rural areas as per 1991 census. These are as follows:

(i) Yield of major agricultural crops (quintals per hectare) (X_1)

(ii) Cropping intensity (X_2), determined by dividing grass cropped area to net cropped area.

(iii) Percentage of irrigation potential created to total cropped area (X_3).

(iv) Yield of major horticultural crops (quintals per hectare) (X_4).

(v) Percentage of families below poverty line to total rural families (X_5) during Eighth Five Year Plan.

Social Indicators

With economic development, social conditions are likely to improve. So the social indicators would certainly point out the levels of economic development/backwardness. Social indicators may be broadly categories as (i) those concerning individual members of the society or groups of members in the society, and (ii) those concerning social services i.e. institutions providing services to the society. Keeping this view, the following social indicators are selected:

(i) Literacy rate (Percentage) (X_6)
(ii) Number of educational institutions per 10,000 population (X_7)
(iii) Number of technical medical personnel per 10,000 population (X_8)

Infrastructural Indicators

The economic development of a backward state like Arunachal Pradesh depends on the development of all sound infrastructural base. The following infrastructural indicators are selected:

(i) Road length per 100 square kilometres (X_9).
(ii) Percentage of villages to total villages having connectivity status of road (X_{10}).
(iii) Percentage of villages electrified to total villages (X_{11}).
(iv) Percentage of villages having drinking water supply to total villages (X_{12}).

The data for the economic, social and infrastructural indicators are collected from the publications of Census of India, 1991, Directorate of Economics and Statistics, Department of Planning and Development, Public Works Department, Irrigation and Flood Control Department, Government of Arunachal Pradesh, together with the unpublished data available with the Directorate of Economics and Statistics and Rural Development Department, Government of Arunachal Pradesh. Most of the data refer to the period of 1996-97 whereas the literacy, population and number of villages data which refers to 1991 census only.

The present study is based on the district as a unit of measurement.

Methodology and Construction of Composite Index

After the selection of indicators or variables and the unit of study, the next step is to consatruct the composite index. The construction of composite index has two important steps namely (a) removing the biasness of scale and (b) determining the weights of the variables.

The biasness of scale may be eliminated by any of the three methods namely (a) by ranking the observation (b) by normalisation of variables and (c) by standarisation of variables.

After eliminating the biasness of scale from their measurement, the crucial problems is that of assigning weights to them. The weights are generally of two kids: (a) subjective and (b) mathematical. In many studies and reports subjective weights are used but they are purely arbitrary and subjectivity entries easily. Hence, it is better to determine weights of different variables with the help of their mathematical properties, thus minimizing the subjectivity in it. A good deal of literature has come up in the last few decades on evolving the mathematical weights as well as composite indices measuring quantitatively the levels of development/backwardness. Among these methods, the most useful and versatile method of construction of a composite index, applicable in many problems of different nature, is the first principal component method an this mathematical method is used in the present study.

Thus more specifically the development indices are calculated by applying Kendall's method of maximising the sum of squared projections (which is statistically identical to the first principal component related to factor analysis), of different districts of Arunachal Pradesh (Kundu, 1980).

The composite index for the different districts of Arunachal Pradesh obtained are as follows:

	District	*Composite Index*
1.	Tawang	2.82343
2.	West Kameng	2.41588
3.	East Kameng	0.27604
4.	Papumpare	1.89759
5.	Lower Subansiri	1.89958
6.	Upper Subansiri	2.00901
7.	West Siang	2.16706

8.	Upper Siang	2.79916
9.	East Siang	3.16459
10.	Dibang Valley	2.92267
11	Lohit	1.81109
12.	Changlang	2.15846
13.	Tirap	2.09847

Classification of Districts

Various techniques of classification of district have been used by researchers and planners. Among them, the quartile method and the average range method are quite popular in this field. These techniques are particularly useful when the number of observation is very large. However, in our case as the number of observation is not very large, so it is easier to classify by taking into consideration the closely related indices i.e. the cluster method under specific groups.

Now on the basis of cluster method the homogeneous districts have been classified into four categories, namely (i) high (H) (ii) Medium (M) (iii) Low (L) (iv) Very low (VL). The district under 'high' category can be considered as relatively 'developed' district. that of medium' category as 'less developed' district that of 'low' category as relatively 'backward' district and that of very low' category as 'relatively more backward' district.

Table 21.1 shows the extent and classification of the levels of development of different districts of Arunachal Pradesh.

Table 21.1: Classification of the Levels of Development of different Districts of Arunachal Pradesh, (1996-97)

Sl. No. *(1)*	*Category* *(2)*	*Name of the Districts* *(3)*
1. High	i)	East Siang (3.1459)
2. Medium	i)	Dibang Valley (2.92267)
	ii)	Twang (2.82343)
	iii)	Upper Siang (2.79916)
	iv)	West Kameng (2.41588)
	v)	West Siang (2.16706)
	vi)	Changlang (2.15846)
	vii)	Tirap (2.09847)
	viii)	Upper Subansiri (2.00901)
3. Low	i)	Lower Subbasiri (I.89958)
	ii)	Papumpara (1.89759)
	iii)	Lohit (I.81109)
4. Very Low	i)	East Kameng (0.27604)

From table 21.1 it is found that East Siang district of Arunachal Pradesh is placed as high category which implies that it is relatively the most developed district among the districts of Arunachal Pradesh. This may be due to the fact that the district enjoys, fertile foothills land and bumper crops are grown. As a result it is found that in terms of the two important agricultural efficiency indicators i.e. yield of major agricultural crops and the percentage of irrigation potential crated to total cropped areas, the position of East Siang district is much higher than that of other districts of Arunachal Pradesh (Appendix 1). Further, it is traditionally a relatively more developed district since the first administrative centre was opened in Pasighat as early as 1911, and as a result the development process started quite early in this district as compared to other districts of Arunachal Pradesh. The East Siang district is also quite advanced in terms of infrastructural indicators. Although the road density of East Siang is below that of the State average, yet we find the highest number of villages (73.68 per cent) having the connectivity status of roads, lies in East Siang district.

The seven districts namely (i) Dibang valley (ii) Tawang (iii) Upper Siang (iv) West Kameng (v) West Siang (vi) Changlang (vii) Upper Subansiri fall under the medium category in the present study. It appears that Dibang Valley and Tawang are well placed in medium category because of their better position in Agricultural efficiency indicator, and to some extent in terms of social indicator, because of their small population base. It is interesting to observe that in case of Dibang valley district the road density is the lowest (6.79 km. per 100 square kilometres) but 44.73 per cent of the villages are having the connectivity status of roads which is nearer to the state average of 47.99 per cent. This may be due to the fact that around 90 per cent of the villages of Roing Circle (where the number of villages are more) are having better connectivity status of roads in spite of having a poor road connectivity circles like Anelhi. Egtalin, Kronli etc. of Dibang Valley district. This calls for a more micro level study i.e. at the block or at the circle level in order to get a better picture, provided the data are available.

The three Districts namely (i) Lower Subansiri (ii) Papumpara and (iii) Lohit are placed in low category. Although Papumpara district is relatively well placed in terms of social and infrastructural indicators (except X_{10} indicator) because of the location of the state capital i.e.,

Itanagar, but it falls under the low category mainly due to its poor performance in the economic indicator, more specifically the agricultural indicator. For example, the data shows that Papumpara district has the lowest yield of major agricultural crops and one of the lowest irrigation potential created to total cropped area, and also the yield of major horticultural crops. It appears that developmental process is concentrate in and around Itanagar and Naharlagun circle of Papumpara district. For example a large number of villages like Kimin, Balijan and Mengio circle of Papumpare district do not have connectivity status of roads. Although in terms of agricultural indicators, Lower Subansiri district is better placed than Papumpare district, but the condition is worse in terms of infrastructural indicators. For example Lower Subansiri district has the lowest percentage of villages having the connectivity status of roads (only 16.30 per cent) and intact more that 90 per cent of the villages do not have connectivity status of roads in Chambang, Damin (Huri) Sarli and Tali circles of Lower Subansiri district. Further Lower Subansiri district has the highest percentage of rural families below the poverty line (78.78%) during the Eighth Five Year Plan. It is Interesting to observe that Lohit district comes under the category of relatively backward district which needs further investigation a more micro level.

Finally East Kameng district of Arunachal Pradesh falls under the very backward district in the present study. This may be because of its poor performance in economic, social and infrastructural indicators i.e., in all categories of indicators, as close look of Appendix I shows that East Kameng district is placed at a much lower ladder as compared to the state average in terms of most of the indicators except X_2 and X_7 indictors because of small population base. Although the sectoral analysis has not been done in the present study but it appears that East Kameng district may need more attention particularly in the field of agriculture and infrastructural base.

In the present study the regional disparity index has been calculated by simple coefficient of variation of composite indices, is found to be 32.27. This may be due to the fact that Arunachal Pradesh is going through a process of transition involving a quantum leap over many intermediate stage of development. The leap is from subsistence economy to the monetised market economy and from isolated villages to urban centres. So, it may be quite natural that the course of development cannot escape from the inverted 'U' Shaped hypotheses, as formulated by Prof. Kuznetsthe renowned development economist, which states that the relative income inequality tends to rise during early stage of development, reaches a peak, and then declines (Kuznets, 1955). However, with economic development the correction of intrastate regional imbalance may not be delayed. Hence the planners are required to chalk out investment planning so that with rapid economic development the regional imbalance within the state may be narrowed down within a defined time horizon and the relatively backward districts catch up with the state average. Thus, it is hoped that the present study will be able to throw some light in isolating areas of need in which State Planning Board will be in a position to reduce inter-district inequalities in future.

REFERENCES

Baster, N (ed) *Measuring Development, the Role and Adequacy of Indicators*. Frank class & Co. Ltd.,1972.

Dasgupta, B. 'Socio-economic classification of Distirct-A Statistical Approach' *Economic and Political Weekly* August 14, 1971.

Kendall, M.G. 'The Geographical Distribution of Crop Productivity in England' *Journal of Royual Statistical Society*. Vol. 1.102,1939.

Kundu, A. *Measurement of Urban Process : A Study in Regionalisation*. Popular Prakashan, Bombay, 1980.

Kuznets, S., Economic Growth and Income Inequality' *American Economic Review*, Vol. 45, March, 1955.

Pal, M.N., 'Regional Disparities in the Levels of Development of India' *Indian Journal of Regional Science*, Vol-II.No.1.1975.

Pathak, C.R.& Kundu, A. Critique of the Techniques for measuring the levels of Development' An abstract Presented to the Symposium in the Regional Disparities in India, New Delhi, 1973.

APPENDIX

Indicators of Development in Arunachal Pradesh

Sl No.	*Districts*	*X1*	*X2*	*X3*	*X4*	*X5*	*X6*	*X7*	*X8*	*X9*	*X10*	*X11*	*X12*
1.	Tawang	12.36	107	46.82	12.61	73.04	29.80	25.81	4.60	37.40	25.66	99.98	73.15
2.	West Kameng	11.56	118	17.45	13.35	25.61	46.31	26.94	4.96	20.06	48.89	89.41	81.96
3.	East Kameng	11.93	127	16.71	0.80	75.73	26.20	24.61	3.97	17.30	29.96	25.45	71.18
4.	Papumpare	8.02	125	12.54	1.14	49.03	55.11	30.55	6.59	23.00	43.02	79.76	76.23
5.	Lower Subansiri	10.85	120	35.32	2.38	78.98	30.07	18.40	3.12	12.30	16.30	72.94	72.94
6.	Upper Subansiri	11.12	191	17.24	5.05	61.61	38.31	28.15	3.79	13.47	30.64	29.19	77.08
7.	West Siang	12.12	102	7.41	10.38	65.14	45.64	27.13	5.89	16.42	45.72	34.06	92.91
8.	Upper Siang	15.53	121	47.78	8.10	63.76	35.33	21.96	9.72	14.14	66.67	18.05	89.48
9.	East Siang	18.52	114	62.54	4.42	53.29	47.85	18.23	10.44	15.92	73.68	89.19	96.38
10.	Dibang Valley	14.48	105	43.83	1.61	34.43	46.88	16.49	9.28	6.79	44.73	38.23	81.43
11.	Lohit	11.48	103	21.43	1.33	52.39	49.21	15.68	3.40	11.91	40.70	41.07	83.77
12.	Changlang	13.01	105	23.34	2.57	21.43	43.20	18.52	2.51	21.50	60.00	85.96	89.82
13.	Tirap	8.10	138	9.31	1.10	67.46	31.06	16.95	5.15	38.27	57.23	96.77	94.23

Note : X_1 X_2 X_3 X_4 X_5 refer to econoomic indicators. X_6 X_7 X_8 refer to social indicator, X_9 X_{10} X_{11} X_{12} refer to Infrastructural Indictors

Source : Compiled from various Departments. Govt. of Arunachal Pradesh which in detail are mentioned in the next.

22

ELDERLY POPULATION OF TRIPURA

— *Paramita Saha*

I

Economic development is accompanied by demographic transition, which involves changes in the age-structure of the population resulting in ageing of the population. Population control strategies directed at reducing fertility and family size induce changes, which are not always socially harmonious or economically benign. Over a period of time shrinking family size will mean the narrowing and perhaps breaking down of old family support systems. Erosion of traditional support systems of family care and security have important bearing on the conditions under which the elderly population live. If the state cannot provide alternative support systems to replace the traditional ones, they will emerge as private enterprises. Market logic will dictate that a large majority of poor people will have no access to these services. Thus it becomes important to device appropriate alternatives of support and care and it becomes necessary to make an assessment of the number and requirement of the elderly people.

This paper aims to present the figures pertaining to the elderly people in Tripura. In section II the process of population ageing in Tripura is described in the context of the national scenario. Section III captures the gender dimension of ageing process in Tripura. Section IV highlights the impact of the ageing process as reflected in the marital status and extended life-span of the elderly. Section V considers work participation of elderly population.

II

Population ageing refers to the changes in age-structure of population and is defined in terms of proportions of persons aged sixty years or more in the total population. In India a steady decline in fertility and mortality rates has accelerated the ageing process.

Fall in fertility results in reduction of share of young people in total population, causing ageing at the base. With fall in fertility levels, the adults of the parental age, of the future, decrease. Thus sustained low levels of birth rates results in reducing future births.

With improvements in the chances of survival, due to better health care, mortality gets reduced and life expectancy increases. As a result share of the older people in the population rises, through ageing at the apex. The contribution of base and apex ageing in the ageing process vary depending on the changes in birth and death rates.

In India the number of elderly is expected to rise from 2.6 million in 1991 to 6.6 million in 2021. While elderly population constituted 6.5 per cent of the total population of the country in 1991, by 2001 the percentage is expected to be 7.63.

United Nations classifies populations with 4 to 7 per cent persons in ages sixty years and above as mature and those with the percentages more than 7 as 'old' or ,aged'. Thus as shown in Table 22.1 Tripura can be considered a state with a population classified as 'old'.

Table 22.1 Elderly Population of Tripura

Year			*Total Population*		*Elderly Population*	
	India	*Tripura*	*India*	*Tirpura*	*India*	*Tirpura*
1961	5.63	5.41	21.51	76.86	26.00	-
1971	5.97	6.69	24.80	36.28	32.32	65.57
1981	6.42	7.05	24.66	31.92	34.50	41.24
1991	6.55	7.01	23.16	34.30	25.73	25.12

Source : Census Reports, 1961, 1971, 1981, 1991.
Note: Age distribution for 1951 was not available.

Decadal percentage variation of total population in Tripura has been much higher than the national figure in all census years from 1961 to 1991. In case of the elderly population, by 1991 the decadal percentage variation in Tripura has fallen below the national figure.

The ageing index i.e. proportion of population above sixty years of age per hundred population, in the 0 to 14 years age-group, has also been increasing as shown in Table 22.2.

Table 22.2 Ageing Index of Tripura

Year	1961	1971	1981	1991
Ageing Index of Tripura	12.64	14.9	17.9	18.3

Source: Census Reports, 1961, 1971, 1981, 1991.

III

In the population composition of India, males have consistently exceeded females in every census period. However, the sex-ratio, measured as proportion of females per thousand males, seems to somewhat improve among the elderly. Table 22.3 shows the sex-ratio among the elderly and total population in the years 1981, 1991 in India and Tripura.

Table 22.3 Sex-Ratio in India and Tripura

	Elderly	*Population*	*Total*	*Population*
Year	*India*	*Tripura*	*India*	*Tripura*
1981	960	921	934	946
1991	930	974	927	945

Source: Census Reports, 1981, 1991.

It is interesting to note that, while for the country as a whole sex-ratio among elderly has sharply declined (from 960 in 1981 to 930 in 1991) in Tripura it has recorded a substantial increase (from 921 in 1981 to 974 in 1991). As for India, in Tripura also, sex-ratio among the elderly is better than the corresponding figure for all age group. This reflects the higher life expectancy of females.

In Tripura the proportion of elderly females in total elderly population, though lower than that of males in total elderly population, has been increasing over time, as shown in Table 22.4.

Table 22.4 Percentaige of Females in Elderly Population of Tripura

Year	1961	1971	1981	1991
Females in Elderly Population	45.83	46.14	47.93	49.33

Source: Census Reports, 1961, 1971, 1981, 1991.

Table 22.5, showing the sex composition of elderly population in rural and urban areas of Tripura throws up an interesting fact. Proportion of elderly females in urban elderly population not only exceed proportion of elderly males in the urban area, it also exceeds the corresponding figure in the rural area, both in 1981 and 1991.

Table 22.5 : Sex Composition of Elderly Population in Rural and Urban Areas of Tripura

	Elderly in Rural Areas		*Elderly in Urban Areas*	
Year	*Male*	*Female*	*Male*	*Female*
1981	52.55	47.44	47.87	52.12
1991	51.43	48.56	46.37	53.62

Source: Census Reports, 1981, 1991 Series 24

Both in urban and rural areas proportions of elderly females have improved from 1981 to 1991. However, proportion of elderly females in rural area has remained lower than that of elderly females in urban area. This is in conformity with national trend of higher sex-ratio in urban areas (India 1997). This may be indicative of (i) availability of better health care and, /or (ii) a more pronounced gender bias working against women in rural areas.

IV

One aspect of the ageing of population is the extension of life-span. Two consequences are particularly important.

1. A prolonged period of multigenerational family life, which need greater social economic and emotional adjustment within the family.
2. The incidence of widowhood, which is an outcome of gender difference in age at marriage, differential life expectancy of males and females, differing proportions of remarriage among elderly males and females.

In a situation when the family is faced with a choice of allocating limited resources, a bias works against age. Even when finances are not the constraining factors, time and effort required for care and support of the elderly, may pose difficult problem for the younger people, who are likely to be engaged in work outside home.

Marital status has a strong effect on the situation of elderly people regarding the availability family care and support.

Table 22.6, highlights the gender difference in the marital status. Most of the elderly males (more than 85 per cent) are married and only a few of them are widower. But majority of female elderly (about two-third of them) and widowed and only slightly more than one-third are married. As expected incidence of divorce or separation is negligible in the older age group, though it shows a slight increase from 1981 to 1991 in all categories.

Table 22.6 : Marital Status of Elderly Population in Tripura

Category	*Married*		*Widowed*		*Divorced/Separated*	
	1981	1991	1981	1991	1981	1991
Male	85.06	86.45	13.45	11.36	0.59	0.61
Female	33.46	38.77	65.93	60.34	0.48	0.50
Total	60.33	62.93	38.60	35.52	0.53	0.55

Source: Socio-Economic Tables. Census of India, 1981, Series 24. Socio-Cultural Tables: Census of India, 1991, Series 24.

The increasing presence of women at the older ages and higher incidence of widowhood among elderly women, have serious economic and social implications. Most of the elderly women do not have the security of maintenance and housing in old age, especially after the death of the husband. High level of illiteracy (Table 22.7), a lack of remunerative occupation as well as lack of awareness about legal and economic rights among elderly women, in comparison with their male counterparts, make elderly women more vulnerable. (Shah, Veedon and Vasi, 1995)

Table 22.7 : Illiteracy among the Elderly in Tripura

Year	*Percentage of Illiterates*	
	Male	*Female*
1981	52.73	89.37
1991	49.44	84.64

Source: Socio-Economic Tables: Census of India, 1981, Series 24. Socio-Cultural Tables: Census of India, 1991, Series 24.

Illiteracy among elderly males and females have recorded a decline from 1981 level to 1991. However, illiteracy among elderly females remains very high at 84.64 per cent.

Though the general practice is to exclude population in the age group 60 years or more, from the working population, it has been found that many of the elderly people continue to work much beyond the working age. They are forced to work to earn a living, as only a few are fortunate enough to have a cushion of pension etc. to see them through the old age. As long as their physical conditions permit they look for employment or other possibilities to earn money. Elderly women, even when they are not working outside home to earn money, they share household chores. As with women of other ages, in case of elderly women also, work within the household goes unnoticed and unrecognised. NSS 42nd Round, which collected data on elderly, reports 89 percent of the elderly women as attending to household duties. NSS also reported that 97 per cent of the men and only 14 per cent of the women were independent.

Table 22.8, summarizes the findings on work-participation of the elderly population on the basis of census data.

Table 22.8 : Work and Elderly Population in Tripura

Percentage of	*1981*	*1991*
Main workers in elderly population	36.73	35.43
Main workers in elderly males	65.00	63.14
Main workers in elderly females	5.70	6.80
Elderly in total main workers	8.70	8.53
Elderly males in total male main workers	9.44	9.28
Elderly females in total female main workers	4.64	4.90

Source : Socio-Economic Tables: Census of India, 1981, Series 24. Economic Tables: Census of India, 1991, Series 24.

The percentage of main workers among elderly males has decreased from 65 per cent in 1981 to 63.14 per cent in 1991, while the percentage of main workers among elderly females has risen from 5.7 per cent in 1981 to 6.8 per cent in 1991. This need not be interpreted to be a consequence of improvement of economic conditions of elderly males. It is rather a reflection. of greater marginalisation of older people in the labour market. This is also supported by the fact that percentage of elderly in total main worker has also recorded a decline from 8.7 per cent in 1981 to 8.53 per cent in 1991. In case of the elderly female the increased proportion of main workers perhaps reflect their economic desperation, which has forced them to greater participation in the labour market. Proportion of elderly females in female main workers has also risen, indicating compulsion on their part to earn more.

In the world of the technology, only the young and resourceful educated and urban, are considered useful. The introduction of technology in traditional rural occupations has left many older people redundant. Increased marginalisation of the elderly, in every sphere of life make them more vulnerable. (Murli Desai, 1999). It is important that society now recognise the problem and wake up to the social responsibility of providing adequate infrastructure for proper care of the growing number of elderly people.

REFERENCES

Census of India: 1961, 1971, 1981, 1991.

Census of India: 1981 Socio Economic Tables, Series-24. 1991 Socio Cultural Tables, Series-24 1991 Economic Tables, Series-24

Dandekar, K (1996): *The Elderly in India*. Sage Publications, New Delhi.

Desai, M: Ageing of Women in Post- Menopausal Stage: Need for Intervention, *EPW* Vol. XXXIV No. 44.

Gulati, L and S I Rajan: The Added Years: Elderly in India and Kerala, *EPW* Vol. XXXIV No. 44.

India (1997): Women in India: A Statistical Profile, Ministry of Human Resource Development, New Delhi.

Karkal, M: Ageing & Women in India, *EPW* Vol. XXXIV No. 44.

Rajan, S I, U S Mishra, PS Sarma (1999): *India's Elderly: Burden or Challenge*, Sage Publications New-Delhi.

Shah, G, R Veedon, S Vasi (1995): Elder Abuse in India, *Journal of Elder Abuse* and *Neglect*, 6 (3/4) 101-18.

Visaria, Pravin (1969): The Sex Ratio of the Population of India, 1961, Vol 1, Monograph No. 10, New Delhi.

23

HOUSING PROBLEMS AND POLICIES OF ASSAM
An Introspection

— *Mouchumi Gogoi Kamelash Kalita*

Introduction

Food, shelter and cloth are the three basic needs of human beings. Shelter means housing provides emotional security and social status of a family. We can study housing as a mental as well as economic activity. As a economic activity it provides income and employment. Indeed, the house is the symbol of its regional character, expressing the distinctive social organisation of its inhabitants, their status and economy. In the view points of rural landscape, the traditional house, forms an outstanding element as an expression of the complex relation between man and his physical milieu. But rapid increased of population, large scale immigration from different parts, haphazard construction on urban as well as rural areas has been increasing the housing problems day to day.

Viewing in mind, an attempt has been made to introspected the problem of housing in Assam.

Objectives of the Study

(a) To study the housing problems and some related housing schemes of Assam.

(b) To give some remedial measures.

Methodology

The data used are collected and analysed from different sources like Govt. and Private Organisations and also field survey and interviews by the researchers.

Discussion and Findings

A. There are three types of houses seen in the different parts of Assam, viz.

(a) Pucca house (R.C.C., Assam Type) 14.62 in 1991.
(b) Semi Pucca House. 15.16 in 1991.
(c) Katcha house or thatched house. 70.22 in 1991.

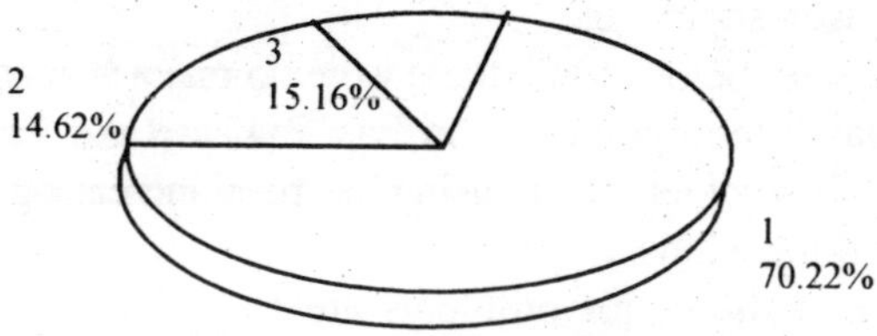

Fig 1 shows in percentage of three types of houses in Assam, 1991

Assam: Number of Households and percentage of households occupying pucca, semi-pucca, katcha, serviceable katcha houses in 1991.

Areas of Census House	*Total no.*	*Percentage of houses in different categories in 1991*				
		Pucca %	*Semi-Pucca %*	*Katcha %*	*Serviceable %Katcha*	*Non-serviceable %Katcha %*
(1)	*(2)*	*(3)*	*(4)*	*(5)*	*(6)*	*(7)*
R	33,26,700	10.53	13.37	76.09	8.79	67.30
URBAN	4,71,660	43.43	27.77	28.80	3.38	25.42
TOTAL	37,98,360	14.62	15.16	70.22	8.12	62.10

Source : Census of India, 1991

Generally, Katcha house are seen in most of the rural areas and pucca and semi pucca houses are seen in most of the urban centres of Assam. Recently, Flat and Apartments are built in Guwahati, Dibrugarh and Jorhat etc.

For well develop houses their need to be some building materials. In Assam building materials are found in three groups :

i. Roof materials - Composed of grass, leaves, wood and other Metal sheets etc.
ii. Wall materials - Grass, mud, or bricks, cements.

iii. Floor materials - Mud floor, wood floor, brick floor, bamboo and cane floors.

C. In respect of housing sites, the oldest villages are found almost invariably on the banks of the rivers and flood plains, but in modern times the suitable sites for the housing has been seen where, better accessibility of transport and communications, and other amenities like, medical, drinking water facility, post and Telecommunications markets, Power supply etc.

D. In Assam, before 1947, there were no major housing problems, but due to large immigration from East Pakistan and Pakistan since 1947 and 1971, problems of housing has been increasing very rapidly in different dimensions.

E. Some of the major problems are :

(i) Some of middle and poor class familie's houses are too small and windows if any, are raised is so small, and no provision of ventilator that scarcely any light and air can penetrate in the room.

(ii) In riverain tracts in Assam, the ordinary cultivator cannot get or can not afford bricks. They are mostly liable to flood, the cultivators live in huts with wetted walls and thatched roof.

(iii) Standard size of housing is a major problem of lower middle classes of family of Assam, only rich family have big sizes of houses.

(iv) Due to increasing population the forest areas are shallowed by men, non availability of building materials and their high costs lead problems to middle classes.

(v) Low holding areas creates congestion problems.

(vi) A major problems is to get sufficient supply of the amenities like sanitation, education, medical, health, drinking water, post and telecommunication, market, power supply etc.

(vii) Floor problem is also one of the problems of Assam. In rainy season kutcha floor become damp and various viral diseases occurred.

There are no any housing policy in Assam except some schemes provided by the organisation like Assam State Housing Board,

Housefed, HUDCO, HDFC, etc. which are generally provided different housing loans under different schemes under different terms and conditions. Some of the major housing schemes provided by these organisation are :

(a) Janata Housing Scheme-.- 24,000 houses for the low economic group of Assam both rural and urban - 25 per cent subsidy.
(b) Composite Housing Scheme:- 1978 - 7000 houses in 27 phases.
(c) Nehru Rozgar Yojana:- 1992 to 1996 are 12,749 houses.
(d) Planted Labours Housing: - Slum clearance Housing Scheme, Subsidize Industrial Housing Scheme on the same line.
(e) Jayanti Village Housing Scheme was an important Housing Scheme for the very poor people of Assam. 500 houses was constructed by this schemes and 100 per cent financed by Govt. of Assam.
(f) Self Finances Scheme is an another important recent (1998) housing scheme with the help of this schemes one can built his house just paying the money to the Housing Board within 6 equal installment.
(g) Regional Building Centre at Bargajai is an pioneer institution under the Assam State Housing Board. It is a training Institute, already 500 youth have trained from this institution for low cost housing construction and building materials production. For example, training production of Hollow Bricks, Ferro cements, production of low cost roofing sheets etc.

Suggestions

(a) To irradicate the urban congestion the social scientist has to inspire and recommend the new houses to built towards the suburbs of the city.
(b) For minimisation of other assessory problems like drinking water, on urban areas one should built their house according to hydrological potential map prepared by geology and mining department, Assam.
(c) Housing provide security and safety, it is a prime need, so

different NGO's and private agencies should come forward for proper implementation of the schemes.

(d) Construction of bamboo and wood floor in the flood havoc zones which can minimize the damage of flood and flood related diseases.

(e) Earthquake prone area should be avoided for construction of tall buildings.

(f) Strict rule of financial schemes should be liberalised.

(g) Environmentally detrimental, scientific, low cost housing technology should be provided, so that all round security and safety should be provided.

In conclusion, it may be said that an alarming and unrestrained consumption of natural resources like land, water, soil, energy, fuel, forest, minerals in human settlements should be conserved and the cause of concern should be given particularly in urban areas. So a housing policy is therefore needed for conserving our resources. The central theme of policy is to creating strong public-private partnership for tackling the housing and habitat issues.

REFERENCES

Nagi, B.H. (1991): *Human Geography*

Pitts & Dickenson : *Introduction to Human Geography*

Rural Development in the Brahmaputra Valley : (1997) A Geographical Analysis *NEIGS*, Vol - 6.

24

DEMAND FOR HOUSING IN URBAN AREA—GAUHATI

— K. Bez

1. Introduction

Like most other urban population of the AFRO-ASIAN Cities and Towns, Gauhatians do share the problems of housing, sanitary facilities, scarecity of drinking water. Yet the growth of urban population increasing at a faster rate, almost 2 out of every 5 persons in the world live in the cities. The drinking water is supplied only to 90 per cent of urban population, although less than 20 per cent is the urban population in India. Availability of sanitation services covered only less than half the population. Migration from rural areas to the towns is understandable, probably more than 60 per cent of GNP is generated in urban areas. They come for a better future. Probably, they do not mind overcrowding, lack of sanitation so long they can earn and save some thing.

Residential Situation

It is reported in Master Plan for Gauhati that as many as 80 per cent of the households with 1-3 persons and 71 per cent of the households with 4-5 persons are rented households. Thus 51 per cent of households with 6 or more persons live in own house. Their survey did not reveal per cent Floor area. It is simply stated that 80, 446 housing units accommodate 80,782 households in the planning area, including old Gauhati Municipality area. The final Master Plan which came out in 1987, showed that as many as 60 per cent of the people live in rented houses. It is disclosed that out of 30,970 households surveyed only 13 per cent are of permanent structures and about 50 per cent semi-permanent and the rest are of temporary structures.

By this time probably most of the temporary and semi-permanent structures are if not demolished declared uninhabitable. At that time the population of planning area was 292,029 which now swelled to more than 10 lakhs, about 244,445 households requiring estimated 204,790 dwelling units. Thus the shortage of households is established. Though a household survey is must to ascertain reliable estimate of demand.

A Theoretical Model

I would like to present a model to estimate the demand for housing units, whether they be owner occupier or rented. we think of a function for the rejected households which may be assumed to be a decreasing one over a specified short time period. This decay function migh assume the form of some established Poisson probability function or weibull probability distribution function. But I did not consider here either of them. Because we have to know the rate of requirements, innovation or condemnation of the households at intervals of time period. A survey means huge cost, which I can not afford. Let us consider a decay model for dwelling units,

$$H = H_o \frac{(H_t - H_o)^{at}}{H^t}$$

$$= H \frac{(1 - H_o)^{at}}{H^t}$$

where

H_t = the no. of dwelling units at time
H_0 = the no. of dwelling units at base period
a = a known constant, say growth rate of unsuitability
and $t = 0, 1, 2, - - - - - - - -T$
For simplicity we write
$H_t = H_c\ (1 - h)^{at}$
with $h = Ho/Ht$
After expasion we find the

$$(1 - h)^{at} = \frac{1 - ath + at\ (at-1)}{2!} \frac{/h^2 at\ (at-1)\ (at-2)\ h^3}{3!}$$

$$= e - ath$$

The expansion beyond the 3rd term is probably not necessary

for practical purposes. Thus, the number of swelling units at time T be

$H_T = H_O\ e\ \text{-}ath$

We also want to estimate the demand for dwelling units not only for those whose dwelling units being condemned but for also additional growth of population.

The immediate demand for dwelling units be estimated from

$$H_t = H_o + (1\text{-}a)\ Ho\ \frac{(H)}{N}\ \text{No.}$$

With $H/N = h$ (say), say the occupancy rate

No = population at the base period whose dwelling units being condemned.

Then we generated this model for further period, thus, we write, with incorporation of population growth,

$$H{=}(H_{O\ e}{}^{-ath})\ \frac{(No\ e^{rt})\ (H)}{N}$$

Where, r = growth rate of population.

The fallibility of this model is apparent, if the time period is longer than 5 years, since it might give explosive demand.

Now we shall consider another model probably more suitable to estimate the demand for owner occupier dwelling units. We take,

$H_t = f\ (N_f\ ,\ P_{tM}\ ,\ P_{tc},\ P_{tc})$

Which may be assumed as functional form of linear economic model.

Where,

N = Population at time t

P_l = Price of land

P_M = Price of building materials

P_C = Price of construction

t = time period

In this model, we dropped the income variable because people would not go to build his own house from running income, the savings may not be declared. Thus, we assume that population, prices of land materials and labour are the explanatory essential variables to estimate the demand for housing.

The Variables

Population : This has already been discussed. The phenomenal growth of city population of 97,384 in 1951 to 693,660 in 1991 which might cross beyond 10 lakh by now.

Price of Land : In the Master Plan area the total land area is 26,177 hectres of which 6,600 hectres i.e. 26 per cent of total area is set for residential purpose, distributed over 9 Zones. It is planned that 6 hectres of residential area for 1000 persons. The land within the Master Plan area would be very costly. The land that costs more or less Rs.75000 in 1980s would now cost probably more than Rs.30 lakhs, yet with zero civic amenities. This is the credit of black money. Thus in estimating the price of land we have to make use of a weighted average formula.

Price of Materal

The architectures would probably be able to give the quantity and volume of material required per square meter. The price will be known from dealers in material.

Cost of Labour

This variable could not be ascertained, there is always variation of cost per square meter from contractor to contractor, also in the case of hired labours. But the variable price could be determined.

Thus, both models could be used for estimating the demand. There are also Regional and District Building centres which might enlighten us further.

REFERENCES

Bez, K. : *Quantitative Techniques in Economics*, Kalyani, 1989.

Bhuyan,D : Future cities : Issues Seminar paper, 1998.

Town & Country Planning Organization Master Plan for Gauhati, 1987.

INDEX